Human 4.0 - From Biology to Cybernetic

Edited by Yves Rybarczyk

Published in London, United Kingdom

IntechOpen

Supporting open minds since 2005

Human 4.0 - From Biology to Cybernetic
http://dx.doi.org/10.5772/intechopen.77612
Edited by Yves Rybarczyk

Contributors
Fatima Isiaka, Julia Bendul, Melanie Zahner, Noé Tits, Thierry Dutoit, Kevin El Haddad, Juliana Alvarez, Pierre-Majorique Léger, Sylvain Senecal, Marc Frédette, Jaime Muñoz-Arteaga, Paolo Baggia, Gabriela Kiryakova, David Brieugne

First published in London, United Kingdom, 2021 by IntechOpen
IntechOpen is the global imprint of INTECHOPEN LIMITED, registered in England and Wales, registration number: 11086078, 5 Princes Gate Court, London, SW7 2QJ, United Kingdom
Printed in Croatia

British Library Cataloguing-in-Publication Data
A catalogue record for this book is available from the British Library

Additional hard and PDF copies can be obtained from orders@intechopen.com

Human 4.0 - From Biology to Cybernetic
Edited by Yves Rybarczyk
p. cm.
Print ISBN 978-1-83880-699-6
Online ISBN 978-1-83880-700-9
eBook (PDF) ISBN 978-1-83880-701-6

Meet the editor

Yves Rybarczyk holds a Ph.D. in robotics from the University of Evry, France. His teaching and research activities focus on artificial intelligence and human-machine interaction. He was an Assistant Professor at the Nova University of Lisbon between 2007 and 2015. Then, he moved to South America where he held the position of Associate Professor and Head of the Intelligent & Interactive Systems Lab at the Universidad de Las Américas, until 2019. Currently, he is a Full Professor and Director of the doctoral programme in microdata analysis at the Dalarna University, Sweden. He has participated in several projects on the modeling and development of complex and interactive systems. He is the author of over 90 scientific publications. Prof. Rybarczyk is a member of the steering committee of the International Summer Workshops on Multimodal Interfaces and an associate editor for the journal IEEE Latin America Transactions.

Contents

Preface

Information technology is becoming ingrained in our everyday life. The consequence of this is that the line between humans and technology is more and more blurred, and tends to transform the human being into a cyber-organism. This transformation, accompanied by the emergence of Industry 4.0, brings us to define a new term: Human 4.0. This new generation of individuals has to deal with smart interconnected pervasive environments supported by the internet of things. Nevertheless, this merge between humans and technology is not straight-forward and requires an additional effort to reduce the gap between the human being and the machine. Such research implies a multidisciplinary approach to the interaction between biological organisms and artificial artefacts. In other words, it is important to perform broader studies from psychological considerations to technological implementations.

To make sure that Human 4.0 will be able to interact with the next generation of semiautonomous systems, it is fundamental to consider both sensorimotor and cognitive properties of the human being. To do so, this book is organised into three sections. The first section comprises studies on the visual aspects of the human-computer interaction (HCI). Chapter 1 analyses the visual aesthetics influencing the behaviour of the end-users. Chapter 2 addresses the question of the immersive power of augmented reality. The second section focuses on sound and speech processing. Chapter 3 presents an overview of the methods used to develop a controllable expressive speech synthesis. Chapter 4 describes the evolution of speech technologies across the last decade. Finally, section three discusses the cognitive treatment of the information in HCI. Chapter 5 consists of a systematic review of the impact of cognitive biases in production planning and control. Chapter 6 exposes enriched measures, such as psychophysiological measurements of the cognitive and emotional state of the end-user, to provide a more advanced assessment of the user experience. To conclude, the last chapter of this volume proposes a model to promote strategies and technological services supporting the collaborative production and access to educational resources.

This book intends to provide the reader with an insight into the new relationship with the technology brought about by the Industry 4.0, and how it can make the human-machine interaction more efficient.

Yves Rybarczyk
Faculty of Data and Information Sciences,
Dalarna University,
Falun, Sweden

Section 1

Visual Processing

Chapter 1

Towards an Optimisation of Visual Aesthetics for User Interaction

Fatima Isiaka

Abstract

Visual aesthetics is a crucial aspect of visual experience, and very few amount of knowledge is distributed to people on how some visual colors are more pronounced than others or why users prefer some colors to others. There are few articles that have written topics on the natural adaptations and how colour can affect people. In this chapter, we lay special emphasis on improving methods on visual aesthetics for user interaction by applying natural valence modelling where color preferences arise from user's average affective response to visual aesthetics, that is mostly related to objects or things around us. A simple experiment conducted as provided support in respect to this phenomenon. Users like or prefer very strong and sharp colors that attract the eyes and dislike colors that are less sharp or clear to human vision. This natural valence modelling agrees more to the data collected and gives a more plausible or very comprehensive meaning to how users prefer the colors of objects they had viewed.

Keywords: visual aesthetics, user interaction, natural valence modelling, optimization, color vision, empirical experiments

1. Introduction

There are very few things that influence human behaviour, which affect their colour preferences as an important aspect of visual experience, such things could be website design, purchase of electronics or home appliances, and buying of personal thing like clothes, shoes and bags. Most people might not understand why they prefer some colors objects (**Figure 1**) to others due to this psychological or behavioural experiences demonstrated by most users when involved in certain activities from most scientific studies [1–3]. Of recent, these explanations have been speculated on how and why users prefer different colours maybe others more preferable than some. Sometimes, colour choices tend to originate from how they are conveyed to humans or users. Most colours attract users at first contact with the eyes while others simple discourage or feel less attractive to approach. Hassenzahl [4] suggested that colour preferences are subconsciously ignited or planted into human memory on a neutral responsive base that are interpreted as some form of support for the biological adaptive theory of colour preferences. The color planted in human memory improves the performance on evolutionary biological phenomena such as how tentative users find some dresses in red more attractive to wear than dresses in brown or green. So to optimise, genetics tune this to behavioural

Figure 1.
Colour objects for background user interaction.

substantial perceptions which resulted in choices of colours for objects rather than colours of their background and free of their original context.

1.1 Objectives

The objectives of this chapter are basically three:

- to exam few literature review on visual aesthetics and its applications;
- to examine the variations of the colour objects as preferred by users based on visual aesthetics quality; and
- to compute the average reaction of users to colour objects they preferred.

2. Related work

In the research paper by [5–7], they described their integrated research approach to the experimental study of emotional user reactions in consideration of the influential and non-influential quality perceptions of the interactive systems. In this chapter, a model is presented that equally defines instrumental and non-instrumental quality perceptions which are the same as the emotional reactions. The interactive features primarily depend on the system properties but also on the user perceptions and context parameters. The main consequence of a users' perception on visual aesthetics of any color object is based on their behaviour or preferences which are centralised components of user experience. The aesthetics aspect of the non-instrument quality is divided into different dimensions which are related to human senses. The visual, auditory and haptic perception are very important in human emotion-technology interaction and these are stated in [8–11]. Visual aesthetics of materials and products is defined as the extent of which the sensory and formal colors and shape attributes of the product that provides positive visual experience for the users. The visual aspects of products have often been stated as most relevant for users aesthetic response as stated by [12–14]. There are various

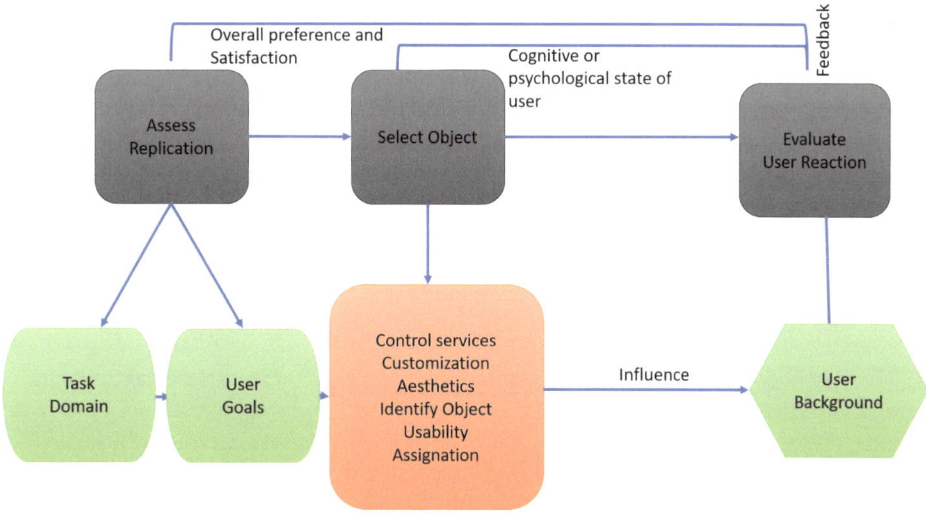

Figure 2.
Approaches to assess users' response to visual aesthetics on objects.

approaches that exist to assess the visual aesthetics of the interactive products presented to users, one of which involve the selected results of two kinds of studies reported in [10], that used the dimension of classical aesthetics proposed by the research work. This is to properly present the relation of perceived visual aesthetics to the perception of the usability, emotional response and consequences of user experience [4, 15, 16]. Research work on overall preference and satisfaction from assess replication to evaluating user reaction (**Figure 2**), as clearly stated steps torwards control service customisation to aesthetics and object identification.

Also, it has been argued that the components associated with usability, such as satisfaction, effectiveness, and efficiency, are never a sole forecast for user experience [17, 18]. The inclusion and assessment of hedonic and pragmatic features regarding user experience technology study about interaction based on pleasure, engagement, pleasure, fun, and games. For mobile devices, it is clear that they have a set of features such as GPS, touch screen and cameras over games consoles that allows for different mode of interactions. Few research as [4, 10, 19–21] stated that aesthetics has demonstrated a positive effect on user perception on mobile system's ease of use, for enjoyment and usefulness. However, the impact of the degree of visual aesthetic quality, which could be high or low on perceived usability is not clear. [22–25] conducted an experiment in the e-commerce domain to explore the hedonic and practical motivation and engagement including both functional and pleasurable aspects of user experience and the hedonic qualities of the colour object properties such as practical and entertaining concepts together with efficiency, functional attributes, cost and influential user engagement. Consequently, there are different kinds of study that have been conducted to assess visual aesthetics and user experience in areas of online news, video games and web surfing. These areas examine the quality of information interactions from users' perspectives. By contrast, perceived aesthetics, usability and focused attention emerged to be typically different from all other visual experience study. Some papers [12, 18] made recommendation on how to make use of fewer constructs in user experience scale on visual aesthetics. Hence, this scale was reduced to four factors which include aesthetics, reward, focused attention and perceived usability. Aesthetics as defined by [19, 20, 26], as the sensuous qualities, the experiences and the emotions that occur while interacting with a colour objects. Visual aesthetics refers to the appearance that user interface depicted, as the top-most visible surface layer of the user experience on user experience modelling [4, 22]. So, perceived usability is defined as the ease of use, goals of efficiency and the practical side of colour object products including satisfaction in specific contextual use of these products. The work we presented tends to use these design goals to shape the way user view objects they observed and how to modify new methods and ways to understand user visual experience on a biologically defined point of view.

3. Methods

The methods used is based on the procedure from [23–25], where participants took part in the experiment for visual aesthetic based on biological association that include warm and cool colours, heavy and light colours, these are pink, blue, red, green, orange and cyan. The ages of these participants are within the ranges of 5–70 years, with equal number of male and female participants. These participants were picked from three institutions, such as Primary, Secondary and Universities. These participants were picked from three institutions, such as Primary, Secondary and Universities. All participants were screened for colour deficiency and none were found to be colour deficient. This was done by placing three main colours (red, blue and yellow) in front of them from the display monitor. All the participants gave consent and the Human Rights Protection Committee (HRPC) of each institude gave approval of the test protocol.

For the task given on colour preferences, emotion ratings and appearances, the participants were shown given 15 colors each (**Figure 3**) with each of them presented one at a time in an arbitrary (random) order. The users described how much they prefer each colour shown to them by specifying reasons for their preferences. They are classified as 'Very much', 'Not likely' and 'Absolutely not' (**Figure 4**) by pressing down on the approval button on the indicators attached to the system and on the display monitor. For each colour, the users rated them as red, pink, blue, light green and cyan. The emotion task is rated as: strong preference, average preference or cool preference and weak preference. For each task the users were allowed five trials each. In addition, the task description was combined to

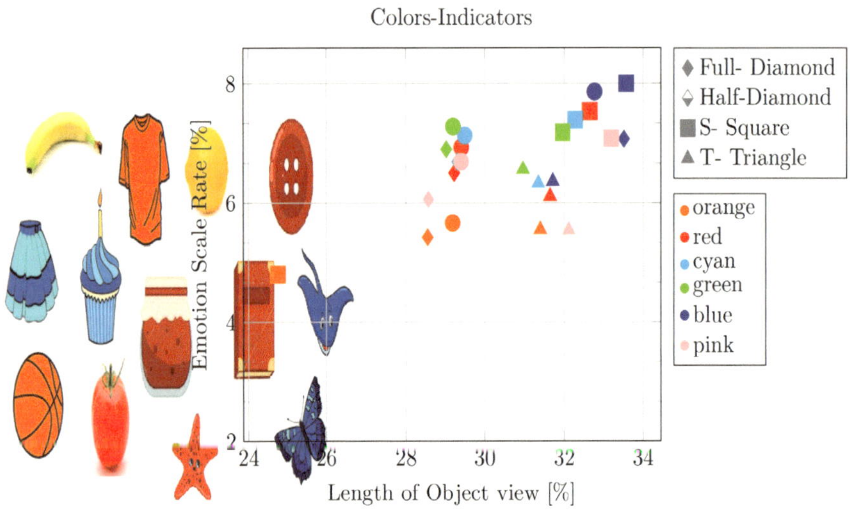

Figure 3.
Colour scale and objects for background user interaction.

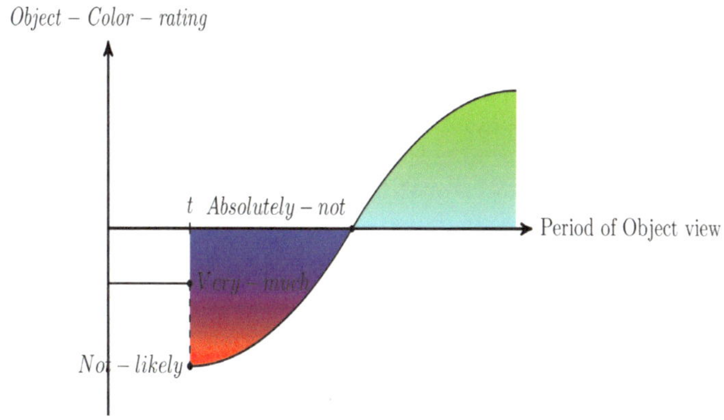

Figure 4.
Colour objects preferences based on biological or cognitive reaction.

superordinate groups if there exist more or less assignment or exemplars that make the same kind of colour object been identified. These colors objects are in form of a bowl, dress, tin, etc., depending on what is been displayed on the monitor. Based on valence rating task, the users were asked to identify each object presented to them on the monitor screen. They are expected to rate their emotional reaction or response to each object they had identified as an answer to the questions been presented, as they select object of certain colour on the monitor screen. That is they are to state how appealing each colour object is to them based on the above classified emotional and assigned description of the objects state above. The objects with similar color were identified and described by the participants individually. Based on this, the users were asked to rate the degree to which each similar color object is to the last one presented on the colour indicator.

4. Results

One of the objectives of this paper is to examine the variation of the colour objects as preferred by users based on visual aesthetics quality, which could be either high or low on these colour objects and also user interface influences perceived colour objects usability. The perception of user data was first inspected for normality based on the equation bellow (Eq. (2)), which is a novel approach, considering the younger participants involved. The error plot did not show any extreme outliers or skewness. For the inverted bell-shaped lines, though slight skewed for red, confirmed that the data were normally distributed. During the preliminary analysis, it was observed that the variability or changes in usability apparent for the shape colour visual aesthetics was higher and apparent than the low or less shape colour objects. The number of participants recruited was not large enough, to some of the user data was simulated to obtain the same user generated data as the original. So the total amount of user data obtained amounts to 60 including that for children. The criteria for detecting and determining the kind of colour object most user preferred by users is based on 'Very likely', 'Not likely' and 'Absolutely not'. Colour objects was utilised because this would attract younger participants and make the result more methodical and generalised.

Based on the computed results above, the error rate for users' visual aesthetics is computed using the equation bellow for both emotion reaction and reasons why users prefer a particular colour object. The reasons why this occur is determined by their colour preferences, which is similar to attributing the colour objects to the once they clicked on to observed. This particular study is more interesting to children, since their psychological reasoning is heightened by the colours and their basic love of colour is exploited. Most of the children preferred rather shape colours like red and Orange while adults prefer frail colours like cyan and blue. The divergence in error is only found in red colour objects because women and few girls tend to go for this type of colours and sometimes some red colour tend to feel less appealing to the eye due to the shade in it (high or low). Some users were mostly confused about their chooses or preferences when it comes to red colour objects. The reason is because the red colour objects rather appear either in light or heavy to the eye when observed by the user. The rest of the colours were measured in the same criteria as the mood of preferences to the user, that is, 'Very likely', 'Not likely', and 'Absolutely not' since they appear to be in-line with the entire colour objects based on the error rate (**Figure 5**).

In regards to the findings, most webpages and games interface can be tuned to these colours as this would likely attract more users and also children would be more drawn to assess such interfaces with average visual console in appearance.

$$U_{pre} = \frac{No - of - User}{100}(1000) + U_{pre} \tag{1}$$

and error is calculated based on:

$$error_{rate} = U_{pre}{}^{1000} \tag{2}$$

The table (**Table 1**) shows the percentage of like and dislike of each colour objects observed by participants based on Eq. (2). This is to reduce the error in computations and predisposition to error in data due the novel approach adopted. The result also showed the variations of usability as well as aesthetics and the predicted impact on the perception of all colour objects (**Figure 6**). Objects with colours associated with a high degree of usability and attractiveness received better ratings than others. The result also showed that emotions of users as regard to the colour objects, revealed that the effect of usability was greater than the one of visual aesthetics for all the valence and emotional or biological feelings involved. Therefore, the colour objects of high regard and appealing design was experienced as most satisfying, while the colour objects of low regard and least attractiveness was most exasperating for children. Both factors contributed to these emotions additively.

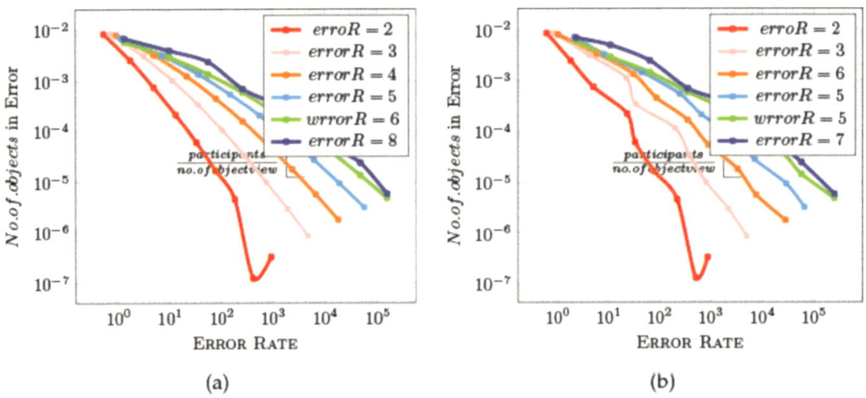

Figure 5.
Graph indicating error in users' emotional rating and colour preferences of each objects they viewed. (a) Error rate based on users' emotional experience, and (b) Error rate based on users' colour preferences.

Colour object	Very likely (%)[a]	Not likely [b] (%)	Absolutely not (%)[c]
Orange	413,450	321,210	4,154,323
Cyan	413,460	391,210	409,324
Red	3,953,430	243,230	406,231
Green	3,965,641	391,210	4,154,323
Yellow	41,345,023	321,210	406,231
Blue	413,451,022	243,230	409,324

[a]*'Very likely' denotes how user prefers the object based on its colour.*
[b]*'Not likely' denotes how user dislike an object based on its colour.*
[c]*'Absolute not' denotes how user totally and not likely to choose an object based on its colour.*

Table 1.
The percentage likes and dislike of colour object as preferred by users.

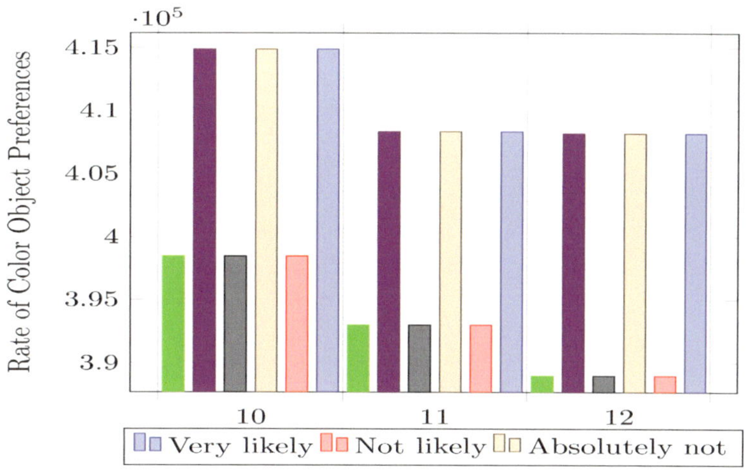

Figure 6.
The Rate of User Color Preferences based on User Perception.

5. Conclusions

The findings in the paper presented, demonstrated that it is possible to compare and deploy groups of colour object properties, which influences high and low regard of colour objects by users. In addition, properties that are associated with statistical presentation had an impact on the perception of usability of colour object properties related to the appearance of the object as determined by users' perceived visual aesthetics. So, it is possible to resolve the issue of attribute overlay and also to establish that high and low colour objects quality perceptions occur autonomously. The paper also showed the relevance of perceived visual aesthetics for emotional user reaction and consequences of the user experience to the object they viewed. The relationship of perceived visual aesthetic and emotional aspects of users have also been studied extensively. Further challenges regarding visual aesthetics in human technology interaction that should be addressed in the future are the role of inter-individual modifications of aesthetics judgements that seem more important, as for example in comparison to the perception of usability issues regarding object colour and the consideration of visual aesthetics in interactive system design projects based on visual colour objects project.

Acknowledgements

The author would like to thank Nasarawa State University, for their support and advise regarding the production of this chapter and choice of the software used for result computation and visualisation.

Notes/thanks/other declarations

Many thanks to my family for their support and contritions to this paper.

Author details

Fatima Isiaka
Department of Computer Science, Nasarawa State University, Keffi, Nigeria

*Address all correspondence to: fatima.isiaka@outlook.com

References

[1] Feijoo C, Gomez-Barroso J-L, Ramos S. An analysis of mobile gaming development: The role of the software platforms. ICIN. 2010; **11-14**:1-7

[2] Ziefle M. Information presentation in small screen devices: The trade-off between visual density and menu foresight. Applied Ergonomics;**41**(6): 719-730

[3] Rogers EM. Diffusion of Innovations. 5th ed. Free Press Trade Paperback; 2003

[4] Hassenzahl M. The interplay of beauty, goodness, and usability in interactive products. Human Computer Interaction. 2004;**19**(4):319-349

[5] Oshita M, Ishikawa H. Gamepad vs. touchscreen: A comparison of action selection interfaces in computer games. In: Proceedings of WASA, Workshop at SIGGRAPH Asia; 2012. pp. 27-31

[6] Bloch PH. Seeking the ideal form—Product design and consumer response. Journal of Marketing. 1995;**59**: 16-29

[7] Bloch PH, Brunel FF, Arnold TJ. Individual differences in the centrality of visual product aesthetics: Concept and measurement. Journal of Consumer Research. 2003;**29**:551-565

[8] Burmester M, Platz A, Rudolph U, Wild B. Aesthetic design—Just an add on. In: Bullinger H-J, Ziegler J, editors. Human Computer Interaction: Ergonomics and User Interfaces. Mahwah, NJ: Lawrence Erlbaum; 1999. pp. 671-675

[9] Desmet P, Hekkert P. The basis of product emotions. In: Green W, Jordan P, editors. Pleasure with Products, Beyond Usability. London: Taylor and Francis; 2002

[10] Schiphorst T. soft(n): Toward a somaesthetics of touch. In: ACM CHI EA '09 Extended Abstracts on Human Factors in Computing Systems; Boston, MA, USA; 4-9 April 2009. pp. 2427-2438

[11] Attfield S, Kazai G, Lalmas M, Piwowarski P. Towards a science of user engagement. In: ACM WSDM'11, Workshop on User Modelling for Web Applications; Hong Kong, China; 9-12 February 2011. pp. (n.d.)

[12] Mahlke S. User Experience of Interaction with Technical Systems. Berlin, Germany: The Technishe UniversitÃ¤t; 2008

[13] Hassenzahl M, Tractinsky N. User experience—A research agenda. Behaviour & Information Technology. 2006;**25**:91-97

[14] Hassenzahl M. The thing and I: Understanding the relationship between user and product; 2003

[15] Hassenzahl M. The effect of perceived hedonic quality on product appealingness. International Journal of Human Computer Interaction. 2001; **13**(4):481-499

[16] Van Schaika P, Ling J. Modelling user experience with web sites: Usability, hedonic value, beauty and goodness. Interacting with Computers. 2008;**20**(3):419-432

[17] Hassenzahl M, Hedonic H. Emotional, and experimental perspectives on product quality. In: Ghaoui C, editor. Encyclopedia of Human Computer Interaction. London: Idea Group; 2006. pp. 266-272

[18] Hassenzahl M. Aesthetics in interactive products: Correlates and consequences of beauty. In: Schifferstein HNJ, Hekkert P, editors.

Product Experience. Amsterdam: Elsevier; 2007

[19] Webster J, Ho H. Audience engagement in multimedia presentations. In: ACM SIGMIS Database: The DATABASE for Advances in Information Systems. Vol. 28, no. 2, 1 Apr 2009. pp. 63-77

[20] Sutcliffe A. Designing for user engagement: Aesthetic and attractive user interfaces. Synthesis Lectures on Human-Centered Informatics. Jan 2009; **2**(1):1-55

[21] O'Brien HL, Toms EG. What is user engagement? A conceptual framework for defining user engagement with technology. Journal of the American Society for Information Science and Technology. Apr 2008;**59**(6):938-955

[22] Jacques RD. The Nature of Engagement and its Role in Hypermedia Evaluation and Design. London: South Bank University; 1996

[23] Quesenbery W. Dimensions of usability: Defining the conversation, driving the process. In: Proceedings of the UPA 2003 Conference; Scottsdale, AZ; 23-27 June 2003

[24] Laurel B. Computers as Theatre. MA, USA: Addison-Wesley; 1993. pp. 1-3

[25] Stephen EP, Karen BS, Jonathan S. Visual aesthetics and human preference. Annual Review of Psychology. 2012. DOI: 10.1146/annurev-psych-120710-100504

[26] Hassenzahl M, Ullrich D. To do or not to do: Differences in user experience and retrospective judgments depending on the presence or absence of instrumental goals. Interacting with Computers. 2007;**19**:429-437

Chapter 2

The Immersive Power of Augmented Reality

Gabriela Kiryakova

Abstract

Augmented reality is one of the technologies that have received great attention and interest in recent years. In a world where the boundaries between the real and the virtual are blurring, this immersive technology enriches and complements the reality with digital content and allows people to gain a complete and real sense of the objects around them. Currently, the applications of augmented reality go beyond the domains of games and entertainment and are aimed at taking the full advantage of the technology in areas such as medicine, architecture, business, tourism, education and more. The current paper presents the essence of technology and types of augmented reality systems. The basic approaches for creating augmented reality applications are discussed. Specific examples of the application of the technology in the field of education are given-an augmented book and augmented reality educational projects, whose purpose is to make learning an interesting, immersive, engaging and motivating process.

Keywords: immersive technology, types of augmented reality systems, augmented books and projects, education

1. Introduction

Technological innovations in the information society have a profoundly transformative impact and significantly change the way people and companies carry out their activities.

According to [1], among the leading technological trends for 2020 is extended reality (XR). This new term encompasses several popular technologies-virtual, augmented and mixed reality-that guarantee immersive experiences. They blur the boundaries between the digital and physical world. The digital world is not just a reflection of the physical environment where people live and work, but it complements the real world with new, more personalized and sensitive sensations. The connection between real and virtual is becoming increasingly narrow and inextricable.

Virtual reality completely immerses users in the digital world, detaching them from the surrounding reality. Augmented reality mixes the real-world physical environments and computer-generated virtual objects and enhances users' perceptions of reality. Mixed reality offers a new environment where physical and virtual objects exist and interact in real time [2].

Augmented reality directly engages and provokes user interaction with both the real-world objects around them and the generated and provided virtual content.

 IntechOpen

Users can use different devices such as smartphones, tablets and wearable devices to view the content that augments physical world.

The world is keenly interested in augmented reality and the technology is expanding into different areas. More and more people, working in the field of education, are setting the question whether it is possible to use augmented reality as an effective tool to realize the ideas of new pedagogical paradigms and engage the learners' attention. This question is provoked by the effect and success of the game Pokemon Go, which is the most striking example of the application of augmented reality. It was the game that made the technology popular and put it in the public spotlight.

Education is constantly changing to keep up with the trends in society. The main reason for changes is not emerging or improved technologies, but the new learners with their specific needs and requirements. Technologies are just tools to create the necessary learning environment where the training can be carried out in the most efficient way.

Augmented and virtual realities create conditions for the realization of the ideas of the immersive learning. Immersive learning supposes that learners acquire new knowledge and skills in an environment rich in sensations, perceptions and emotions. The major benefits of augmented reality are connection and interaction between all participants in the learning process, physical objects and virtual content.

Increasingly, augmented reality-based educational applications are being offered on the educational application market. Their goal is to engage learners' attention, enhance their participation in training and help the perception of abstract scientific concepts by specifying and connecting them to the real world.

2. Augmented reality technology

Augmented reality plays the role of the bridge that connects the digital and real world. It does not replace the physical world, as virtual reality does, but complements, expands and enriches the real physical environment with layers of computer-generated content. The technology provides interactive, accessible and digitally manipulative information about real objects. It transforms the way people interact with virtual content and creates the preconditions for engaging all human senses in perceiving the physical world.

Ref. [3] defines augmented reality as a system that:

- combines real and virtual objects. In augmented reality systems, virtual and real objects coexist at the same time, in the same place.
- offers real-time interactivity. Users can interact with virtual content, which responds to their actions.
- allows registration of virtual objects in the physical 3D world. There is a geometric alignment of virtual objects with real ones in the physical world.

Augmented reality enhances users' perception of and interaction with the real world. With technology, any real object can be enriched with additional information, which sometimes cannot be detected and perceived directly through the users' senses.

The term augmented reality was introduced in 1990 by Tom Caudell. The technology has already had a long history. In 1968, Ivan Sutherland invented the first head-mounted display model, which, despite its weight, is a step toward the use of augmented reality in the real world. In 1992, Luis Rosenberg created the first functioning augmented reality system Virtual Fixtures, designed to train pilots in the US Air Force. The commercial application of the technology dates back to 1998 with the introduction of yellow first down line. In 1999, Hirokazu Kato created ARToolKit-a library for developing augmented reality applications. In 2014, the launch of Google Glass set the era of the wearable devices, which gave a strong impetus to the development and spread of augmented reality technology.

2.1 Augmented and virtual reality: common goal, different approaches

The reality-virtuality continuum, proposed by Paul Milgram, is a continuous scale, varying between reality (a physical space with no virtual elements) and virtuality (a virtual space with no physical elements). The area between the two endpoints is mixed reality-a space where real and virtual objects are combined in different ratio. It is clear from the scale (**Figure 1**) that augmented reality is closer to the reality than to virtuality, since it is a physical space with integrated virtual elements.

When discussing augmented reality technology, analogies are invariably made with virtual reality. The two technologies have a common purpose-to immerse users and give them a complete and real sense of the world around them, but they achieve the goal in different ways.

Virtual reality replaces physical world with a digitally created one. Users are completely immersed in the artificial environment and cannot see, feel or interact with the physical world around them. Virtual reality is one of the technologies ensuring that people are completely focused on situations in the digital world. But the technology detaches them from the physical world and is not possible to make a connection between what is happening in the digital world and the surrounding reality.

Unlike virtual reality, augmented reality creates virtual objects that overlay the physical world and provide additional details for real objects. It allows users to be in touch with the real world and interact with real objects and people as well as with computer-generated content. The basic idea behind augmented reality is to immerse users in the physical world and give them more information about what they see, but are not able to perceive through their own senses.

2.2 Types of augmented reality

There are different classifications of augmented reality according to various criteria.

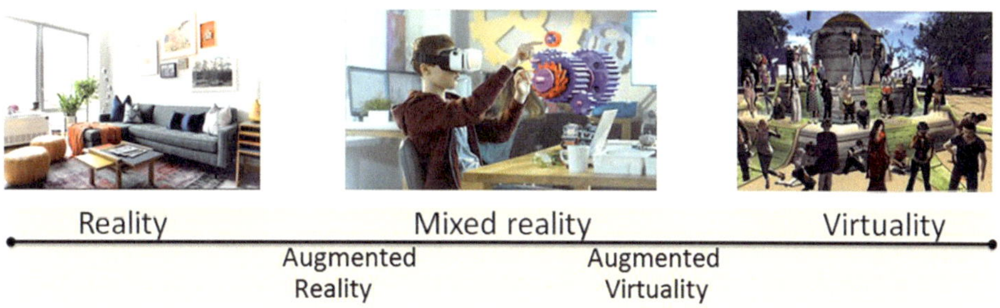

Figure 1.
Reality-virtuality continuum (adapted from [4]).

Depending on the hardware devices that users use, there are the following (based on [5]):

- **Stationary augmented reality systems**. They are equipped with powerful cameras and can provide accurate recognition of objects and scenes from the reality. Examples of stationary systems are the virtual fitting rooms created in physical stores.

- **Spatial augmented reality systems**. They are also known as video mapping or projection mapping. Spatial systems project virtual content in actual size and proportions directly onto real-world objects. Many companies from automotive industry use such systems to present their new models. On the other side, video mapping show programs are receiving great attention from consumers.

- **Desktop augmented reality systems**. These systems use the device camera to recognize objects or scenes from the reality and embed virtual objects, both displayed on the desktop screen. Many companies use desktop augmented reality to create virtual fitting rooms in their online stores.

- **Mobile augmented reality**. This type of augmented reality is extremely popular recently due to the widespread of mobile devices (phones, tablets and hand-held devices). Many mobile augmented reality applications use the global positioning system (GPS) capabilities of the devices to determine users' position and provide and show information that is directly related to their location. The information comes in different formats (text, images, audio and video) and is integrated into the real environment.

- **Head-mounted displays**. This group of hardware devices delivers a combined image of the physical world and virtual objects. Depending on the used technology, head-mounted displays can be divided into optical and video see-through devices. The most popular representative of this group of hardware devices is smart glasses.

- **Contact lenses**. They are the future of augmented reality. Although they are still under the development process, Bionic Lens is already being tested.

Depending on how the objects or scenes from the real world are identified, the augmented reality is divided into:

- **Vision-based augmented reality**. The augmented reality applications recognize and interpret real-world images, scenes or objects. Vision-based augmented reality can be marker-based or markerless. Marker-based applications use the camera of the device to recognize and interpret markers, usually black and white barcodes. The software analyzes the marker and creates virtual objects that are displayed on the device screen integrated to the recognized real objects. Markerless technology uses real-world objects (photos, objects, scenes, etc.) as targets and the software recognizes them by specific features that distinguish them from the surrounding environment.

- **Location-based augmented reality**. It is a kind of markerless augmented reality. Applications use the GPS capabilities of the devices to determine their position and provide content that is relevant to the current location of the users-for example, information about hotels, restaurants, museums and others that are close to the users.

2.3 Augmented reality systems work process

There are four main stages in augmented reality systems' functioning [5]: scene capture, identification, processing and preview.

The reality that has to be augmented with digital content is captured by video or optical see-through devices. The captured reality is scanned to determine the position of virtual content that augments it by markers or tracking technologies. When the scene becomes recognized and identified, relevant virtual content is required from different sources. The final step is to create a mixed image of the real scene and embedded virtual content.

Users need the appropriate software to view and interact with digital content that augments the physical world. They can use augmented reality browsers or specific applications.

3. Augmented reality: an innovative learning tool

Recently, augmented reality is a popular tool used in various fields of business, advertising and tourism. There are many examples of successful augmented reality-based campaigns. Some companies choose this technological approach to increase their sales, attract more customers and guarantee their social presence. Others use augmented reality to create a personal and emotional connection between the products and services they offered and potential and actual customers. Some companies rely on unforgettable authentic experiences that can be offered to consumers through augmented reality applications and subsequent positive results from such campaigns.

Augmented reality can improve the perception of the surrounding real world with new sensations and perceptions, which is a prerequisite for a better understanding of the physical world [6]. This is of utmost importance for the education sector. Augmented reality makes possible learning in the real world and changes the way learners interact with the objects around them.

Present-day digital learners, on the other hand, are keen to use technologies from their daily lives in the learning process. Like any technology that is able to engage learners' attention and provoke their interest and activity, augmented reality is a subject of interest for teachers.

The main questions that should be asked in an emerging idea of implementing new technologies in education are as follows:

- Why to use them and what positive effects can be expected?
- How they can be used to support the realization of dominant pedagogical paradigms?

3.1 Why to use augmented reality in education?

The concept of immersive learning has become more and more relevant in recent years. Immersive learning is a theory based on the use of technologies and technical tools to create an educational environment, where learners are immersed in new worlds, learn through different perceptions and senses and have first-person experiences [7]. The idea is to place learners in new situations or environments that offer different sensations and experiences and create conditions for more curious and motivating learning. Technologies that help realize the ideas of immersive learning are virtual reality and augmented reality.

With augmented reality, the training is carried out in a different type of learning environment, with opportunities to enrich the learning content and acquire new skills and competences.

The technology creates conditions for building a new type of learning environment that combines reality and the virtual world. The boundaries between them are blurred and a united learning space is created that is rich in possibilities, perceptions and sensations and offers new ways of acquiring knowledge and skills [7]. Learners are placed in an authentic learning environment and provided with a real-world experience [8, 9]. Every real object is enriched with dynamically changing information. Learners may receive additional information that depends on their specific actions with objects and each interaction can cause a different reaction, which can be a source of new knowledge for them [10]. The interaction with real and virtual objects is a way to realize active learning.

Learners can interact, explore and experiment with virtual objects, discovering their properties and behavior-something that is not always possible in the real world or is not safe. There are many examples of using augmented reality in training-from studying the structure of molecules and atoms, exploring the universe, the planets and the Earth, to acquiring knowledge for the structure of the human body and the functioning of human organs and systems. Learners are active participants who take control of the interaction with digital content in real time. They have the ability to take decisions and actions in order to achieve the learning objectives [7]. Such kind of learning approach can facilitate the understanding of abstract or complex concepts through their concretization [11, 12]. At the same time, it provokes imagination and creativity and increases the learners' interest and motivation.

In a hybrid learning space, learners can interact not only with the real objects and the digital content that augments them but also with other participants [7]. The learning environment allows the implementation of some of the approaches of modern smart pedagogy, where learners are active, motivated and engaged participants-learning by doing, applying a research approach to training and working in a constructivist environment, context-based learning.

Today's learners are digital learners. They were born and live with technologies and prefer to study in the digital environments. Most of them spend more and more time using various electronic devices mainly for chat or games. A serious problem with the reluctance of young people to read books and textbooks is arising. Parents and teachers face difficult questions about how to encourage teens to read more. Integrating augmented reality to traditional books can help solve the problem and generate engagement and fun in reading.

The integration of augmented reality technology and traditional books creates the so-called augmented books. They look like traditional books, but when readers point their smartphones or tablets to pages, the content comes to life. 3D models, video, audio, animations and interactive elements appear. Through technology, print materials are enriched with digital multimedia information, which can make learning content easier to read and more engaging.

Augmented reality supports the acquisition of practical skills and experience through training in a mixed environment and simulations. To be prepared for the real world, learners must have not only knowledge but also practical skills and experience. Augmented reality applications allow them to actively develop their skills by working, manipulating and experimenting with both virtual objects and 3D models of real objects. Augmented reality helps different situations and problems of practice to be recreated, which resolved by learners leads to the acquisition of practical skills and experience.

The ability to manage their own learning stimulates and motivates learners. They can discover new properties and behavior of objects and processes, following their individual research approach. Augmented reality is a technology that is able to inspire learners, spark their interest, stimulate their creativity and curiosity, create positive emotions and attitude to the learning, and enhance their motivation. These are the necessary prerequisites for a positive change in the educational process and its orientation toward digital learners' needs.

Augmented reality is a promising technology that has its place in education. The listed benefits make it a preferred tool for both teachers and learners. The technology enables innovative forms of training and learning, and supports the realization of new pedagogical paradigms.

Augmented reality also has some disadvantages that must be taken into account in order to use technology in the best possible way in education. The technology requires the availability of smart devices, which can put learners in an unequal position. Technical problems of different nature are possible, caused by device camera malfunctions, lack of high-speed Internet connection as well as software issues. On the other hand, it is possible that augmented reality applications may distract and divert the learners' attention from the learning materials and cause undesirable reactions and results. Developing appropriate educational applications is a difficult process. It takes time and resources and requires an innovative approach to present virtual content as well as to choose tools and approaches to access and interact with it.

3.2 How to use augmented reality in education?

Augmented reality can be integrated in different ways in training. It can be used to create augmented textbooks or to develop augmented reality educational applications.

3.2.1 Augmented textbooks

Nowadays, interactive books are very popular because they make reading an active process. Augmented books are a subtype of interactive books.

Augmented book is a traditional (paper) book with additional virtual content, which is accessible through appropriate hardware and software tools [13]. The augmented books enhance the traditional ones with interactive visualizations, animations, 3D models, and audio and video materials. All these multimedia components improve reading and engage readers' attention [14]. A very important feature of augmented textbooks is that they allow learners to interact with virtual content, explore and make experiments, which guarantee active learning, better and easier perception and understanding of complex and abstract concepts. They can have a positive impact on learners' achievements and attitude toward their own learning [15].

Augmented textbooks have a significant advantage over other e-textbooks. People can read them as traditional books, but at the same time, they have an access to additional interactive digital content. There is no need to buy special equipment to use them since a smartphone or tablet is required to visualize integrated digital content-devices that learners use every day.

Augmented reality gives new dimensions to traditional textbooks, which many digital learners consider being boring. The technology can turn textbooks into an interactive tool that engages learners, motivates them to read and learn, as well as entertains them. Much more learning content can be included in augmented textbooks that exceeds the requirements of the curriculum. The additional content is accessible to anyone who has the need or desire to deepen his knowledge.

Figure 2 shows an augmented book Bugs: Interact with Augmented Reality Creepy Crawlies (by Hannah Wilson, Carlton Kids Publisher). After scanning the pages, readers can see 3D models of different species of bugs and interact with them. Magic Books, Animal Kingdom Education Book, Animals Encyclopedia, Secrets of the Ocean, iStorm: Wild Weather and Other Forces of Nature, Dinosaur 4D +, Live Solar System and many others are good examples of books and textbooks that are created with augmented reality technology. Many of them target younger learners to provoke and stimulate their desire to read. Among the augmented textbooks that are used in secondary schools, colleges and universities are the statistical software for processing experimental data, cutting tools, Ethnobotany Workbook, Imagina Books: Human Body and others.

3.2.2 Augmented reality applications

Another approach to use augmented reality in education is to develop augmented reality applications.

More and more products created with augmented reality technology are available in the educational applications market. Many teachers choose such applications as an effective and powerful tool to help students better understand abstract concepts.

Augmented reality can represent objects and processes that are difficult to imagine and turn them into 3D models allowing learners to interact with them. The abstract concepts come to life and become part of the real world, which make them easier to learn.

Augmented reality can be used as an effective tool that "translates" theoretical scientific concepts into things from the real world and daily life, making them understandable to learners.

Physics, astronomy, chemistry, biology and mathematics are among the subjects where augmented reality technology can be used effectively. The various physical and chemical properties of the objects and the existing regularities can be easily understood with the help of augmented reality. The anatomy and structure of the human body and the functioning of human systems can be visualized by augmented reality in 3D space in biology courses. The properties of curves, surfaces

Figure 2.
Augmented book bugs: Interact with augmented reality creepy crawlies.

and 2D and 3D geometric figures can be studied in mathematics and geometry classes. Trips to the farthest corners of the earth and the universe are possible, as well as going back to the past and tracking historical events and places.

Working with virtual objects can help learners to acquire and develop practical skills and experience in safe environment.

Figure 3 shows an educational application with augmented reality-Anatomy 4D that can be used to study human organs and systems. There are markers that need to be printed in advance and 3D models of human organs and systems are displayed after scanning them. The application allows learners to interact with 3D objects and digital content.

The combination of virtual and augmented reality can help to develop educational applications that immerse learners into new worlds with many different sensations. A good example of mixture of both technologies is Google Expeditions.

Google Expeditions is a learning application that allows learners to get to know the world, its history and culture through virtual reality or augmented reality tours.

In virtual reality tours, participants in the learning process use their mobile devices and Google Cardboard to immerse themselves in the digital world and explore the world, the universe, oceans, mountains, caves, galleries and museums without leaving the classroom (**Figure 4**). In augmented reality tours, virtual objects are integrated into the physical environment and they become part of the learning space. Training occurs in a shared environment, where interactions and experiments are possible under the guidance of teachers.

Augmented reality applications generate a great interest and desire to work, because they allow learners to be active, to be researchers and to control their own learning.

Regardless of the chosen approach of integration of augmented reality in learning, the technology creates prerequisites and conditions for acquiring new knowledge and skills in an environment that is rich in sensations, perceptions and emotions.

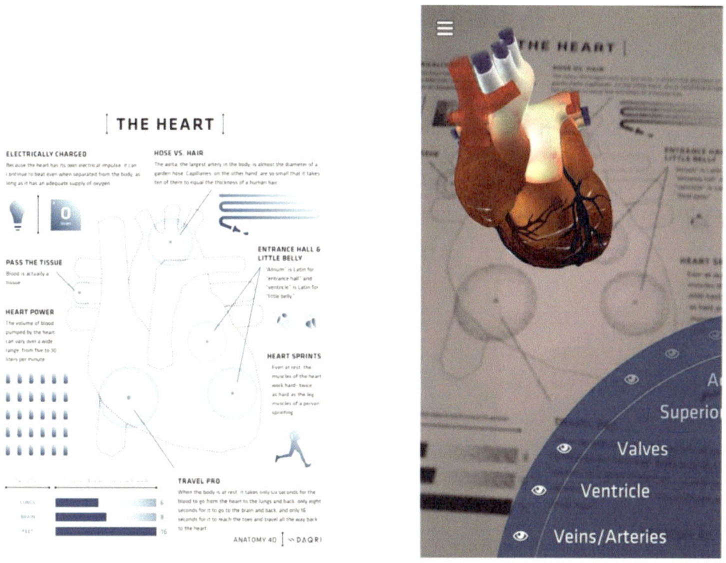

Figure 3.
Educational application anatomy 4D.

Figure 4.
Virtual reality tour with Google expeditions.

4. Tools for creating augmented reality educational applications

There are many augmented reality applications that are specifically designed for education and are free for use for both teachers and learners. Often, teachers prefer to develop their own learning content and materials that are designed and tailored to their learners' goals.

Users can create augmented reality applications with various software platforms (many of them are free to use). There are two main approaches: using augmented reality browsers or augmented reality software development kit.

4.1 Augmented reality browsers

Augmented reality browsers are applications that display virtual geolocation multimedia content upon real-world objects. Browsers access remote resources through web protocols and services, index content through media streams and support various MIME formats [16]. Augmented reality browsers offer not only opportunities for publishing content that augments the reality but also tools for developing augmented reality projects.

With browsers, the process of creating augmented reality projects is quick and easy, and no special programming skills are required since templates are used and most of the procedures are automated. There are some limitations of created projects-recognition of only 2D images is possible and inability for interaction between users and virtual content. A major drawback of this approach is that there are no standards that allow augmented content to be interchangeable. Content created with the tools of one browser cannot be accessed through another browser [17].

Popular augmented reality browsers and their project creation tools include HP Reveal (HP Reveal Studio), BlippAR (Blippar Studio and Blipp-Builder), Zappar (ZapWorks Studio and ZapWorks Designer), Wikitude (Wikitude Studio) and more.

Different terms are used to refer to the created products (layers, channels and worlds), but the principle of action is identical-consecutive steps: (1) selection of the images-targets, which will be scanned; (2) add different types of digital content that will enrich the images; (3) review and test and (4) publish the project. After creating and publishing augmented reality projects, users (learners) must scan the images with augmented reality browser in order to visualize the digital content.

Figure 5 shows an augmented reality project created with Wikitude Studio. When scanning the image (a table with logical functions), learners can see a video material, which illustrates the results of the logical functions and their dependence on the values of the input arguments. Video material is created by Google Doodle and is accessible in YouTube.

4.2 Augmented reality software development kit

Augmented reality software development kits (AR SDK) have fundamental features such as image, object and location recognition, position estimation and others. Among the popular AR SDKs are ARKit, ARCore, Vuforia Engine, Wikitude AR SDK, Kudan, AR SDK, EasyAR SDK and ARToolKit. They all allow 2D image recognition and are compatible with different platforms (iOS, Android, Windows and others). Some of them provide 3D object recognition features, while others help create geolocation applications. AR SDKs can be used to create augmented reality applications for smartphones, tablets and a variety of wearable devices. This approach results in developing stand-alone applications that users use to view virtual content that augments and enriches real objects.

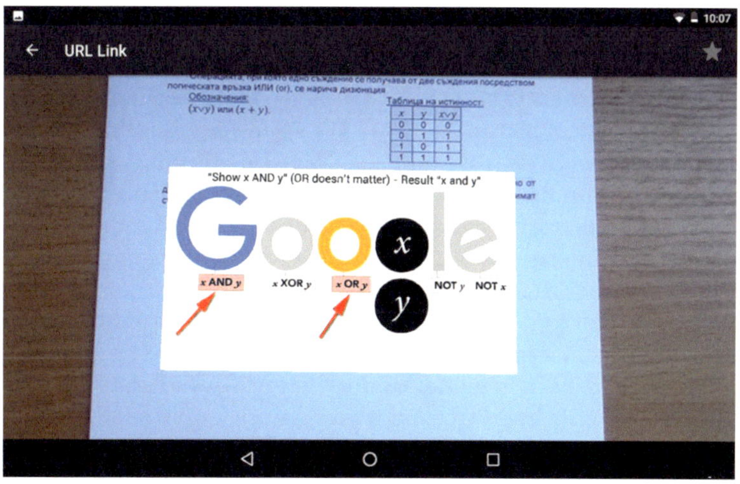

Figure 5.
Wikitude studio project.

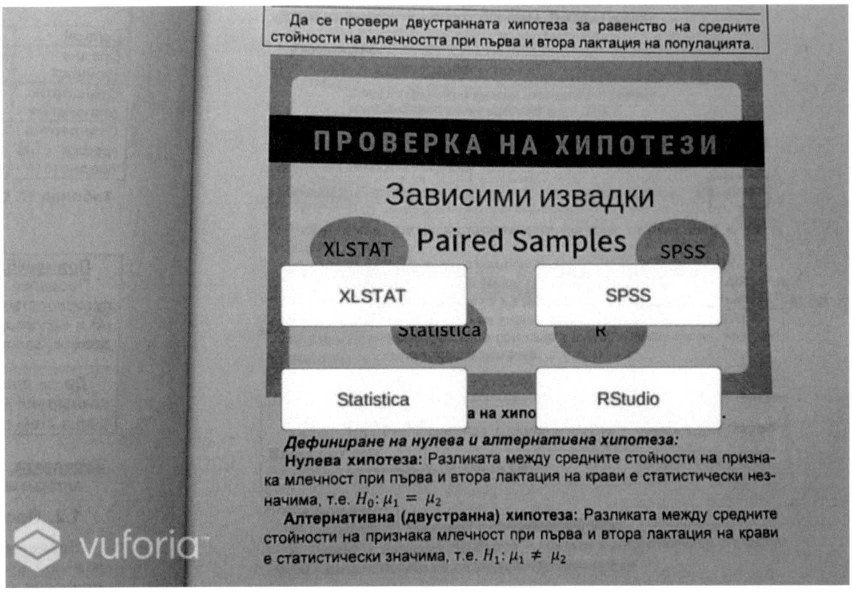

Figure 6.
Virtual buttons in BookStatistica application.

Augmented reality applications, created with AR SDKs, are interactive, are feature-rich and embody the best of augmented reality technology. Their development is a complex and time-consuming process, requiring programming skills.

Figure 6 shows an augmented reality application BookStatistica, developed with Unity and Vuforia Engine. After scanning different markers, virtual buttons appear on the device screen. Learners can make a choice between different options-for example, video materials that show how statistical methods are done in different statistical packages.

5. Conclusion

The digital transformation is inherent in today's society and is possible due to the diversity of information and communication technologies and comprehensive Internet access. The transition covers all spheres of people's lives and is rapidly advancing into education. It is compulsory for the educational goals to incorporate ideas for preparing learners to work with digital technologies in the digital world.

Technologies in education can greatly facilitate all activities in the learning process. There are different ways to integrate them into training. They can be used to deliver traditional training and guarantee automating and speeding up of most of learning and administrative activities. But the most important idea behind the use of innovative technologies is to change the role of learners-to put them in the center of learning and turn them into active participants, who construct their own knowledge, follow their personalized learning path, and create and share content with other learners in an interactive mode.

Today's learners are born in digital world. They use different devices and technologies in their daily lives but cannot successfully implement them in an educational environment [7]. Therefore, the teachers' part is very important. They have to integrate technologies in the most appropriate way in the training, taking into account the needs and characteristics of learners in order to maximize their educational benefits.

Augmented and virtual realities have a great potential in the field of education because they can significantly change the way the learning is carried out and fill it with new sensations and perceptions. Augmented reality creates an immersive educational space that is a mix of real world and digital content and offers rich experience. Learners can work and interact with virtual objects that are integrated into the real world. They receive context-based digital information for each observed physical object or process. The use of the technology makes it possible for learners to understand complex and abstract theories and concepts much faster and more easily that inspires and motivates them. Augmented reality is one of the innovative technologies that combine attractiveness and effectiveness of the training. It brings to life print textbooks and turns them into attractive, fun and interactive tools that make reading and studying active processes. A significant advantage of augmented reality to virtual reality is that it overcomes the risk of learners' social exclusion and insufficient social and communication skills since they are not isolated in virtual worlds and can interact with other learners.

Augmented reality is among the innovative technologies that are able to transform education into smart education to correspond to the developing smart digital society.

Conflict of interest

The author declares no conflict of interest.

Author details

Gabriela Kiryakova
Faculty of Education, Trakia University, Stara Zagora, Bulgaria

*Address all correspondence to: gabriela.kiryakova@trakia-uni.bg

References

[1] Marr B. The 7 Biggest Technology Trends In 2020 Everyone Must Get Ready For Now [Internet]. 2019. Available from: https://www.forbes.com/sites/bernardmarr/2019/09/30/the-7-biggest-technology-trends-in-2020-everyone-must-get-ready-for-now/#25b725042261 [Accessed: 04 December 2019]

[2] Wikipedia [Internet]. Available from: https://www.wikipedia.org/ [Accessed: 04 December 2019]

[3] Azuma RT. A survey of augmented reality. Presence Teleoperators and Virtual Environments. 1997;**6**(4):355-385

[4] Milgram P, Kishino F. A taxonomy of mixed reality visual displays. IEICE Transactions on Information and Systems. 1994;**77**(12):1321-1329

[5] Glockner H, Jannek K, Mahn J, Theis B. Augmented Reality in Logistics. Troisdorf, Germany: DHL Customer Solutions & Innovation; 2014

[6] Kesim M, Ozarslan Y. Augmented reality in education: Current technologies and the potential for education. Procedia-Social and Behavioral Sciences. 2012;**47**:297-302

[7] Martín-Gutiérrez J, Mora CE, Añorbe-Díaz B, González-Marrero A. Virtual technologies trends in education. EURASIA Journal of Mathematics, Science and Technology Education. 2017;**13**(2):469-486

[8] Hwang GJ. Definition, framework and research issues of smart learning environments-a context-aware ubiquitous learning perspective. Smart Learning Environments. 2014;**1**(1):4

[9] Zhu ZT, Yu MH, Riezebos P. A research framework of smart education. Smart Learning Environments. 2016;**3**(1):4

[10] Billinghurst M. Augmented reality in education. New Horizons for Learning. 2002;**12**(5):1-5

[11] Lee K. Augmented reality in education and training. TechTrends. 2012;**56**(2):13-21

[12] Yilmaz RM. Augmented reality trends in education between 2016 and 2017 years. State of the Art Virtual Reality and Augmented Reality Knowhow. 2018;**81**:97

[13] Altinpulluk H, Kesim M. The classification of augmented reality books: A literature review. In: Proceedings of the INTED. 2016. pp. 4110-4118

[14] Dünser A, Walker L, Horner H, Bentall D. Creating interactive physics education books with augmented reality. In: Proceedings of the 24th Australian Computer-Human Interaction Conference. ACM; 2012. pp. 107-114

[15] Lim C, Park T. Exploring the educational use of an augmented reality books. In: Proceedings of the Annual Convention of the Association for Educational Communications and Technology. 2011. pp. 172-182

[16] Grubert J, Langlotz T, Grasset R. Augmented reality browser survey. Technical report. Austria: Institute for Computer Graphics and Vision, Graz University of Technology; 2011

[17] Ogden H. An Introduction to AR Browsers [Internet]. 2015. Available from: https://ercim-news.ercim.eu/en103/special/an-introduction-to-ar-browsers [Accessed: 04 December 2019]

Section 2

Sound Processing

Chapter 3

The Theory behind Controllable Expressive Speech Synthesis: A Cross-Disciplinary Approach

Noé Tits, Kevin El Haddad and Thierry Dutoit

Abstract

As part of the Human-Computer Interaction field, Expressive speech synthesis is a very rich domain as it requires knowledge in areas such as machine learning, signal processing, sociology, and psychology. In this chapter, we will focus mostly on the technical side. From the recording of expressive speech to its modeling, the reader will have an overview of the main paradigms used in this field, through some of the most prominent systems and methods. We explain how speech can be represented and encoded with audio features. We present a history of the main methods of Text-to-Speech synthesis: concatenative, parametric and statistical parametric speech synthesis. Finally, we focus on the last one, with the last techniques modeling Text-to-Speech synthesis as a sequence-to-sequence problem. This enables the use of Deep Learning blocks such as Convolutional and Recurrent Neural Networks as well as Attention Mechanism. The last part of the chapter intends to assemble the different aspects of the theory and summarize the concepts.

Keywords: deep learning, speech synthesis, TTS, expressive speech, emotion

1. Introduction

Controllable Expressive Speech Synthesis is the task of generating expressive speech from a text with control on prosodic features.

This task is positioned in the emerging field of affective computing and more particularly at the intersection of three disciplines:

- Expressive speech analysis (Section 2), which provides mathematical tools to extract useful characteristics from speech depending on the task to perform. Speech is seen as a signal, such as images, text, videos, or any kind of information coming from any source. As such, it can be characterized by a time series of features.

- Expressive speech modeling (Section 3), modeling human emotions and their impact on the speech signal. Speech is considered here as a means of communication between humans.

- Expressive speech synthesis (Section 4), for which machine learning tools have become ubiquitous, especially hidden Markov models (HMMs) and more recently Deep Neural Networks (DNNs). The field of Machine Learning allows machines to learn solving a given task. This field borrows from an ensemble of statistical models allowing to represent or transform data. It also uses concepts from Information Theory to measure distances between probability distributions.

2. Expressive speech analysis

2.1 Digital signal processing

A signal is a variation of a physical quantity carrying information. The acoustic speech signal is converted into an electrical signal by a microphone. An acoustic signal is a variation of pressure in a fluid that the human perceives through the sense of hearing. This signal is mono-dimensional because it can be represented by a mathematical function with a single variable: pressure.

The electrical signal generated by the microphone is an analog signal. In order to process it with a digital machine, it must be digitized. This is done by electronic systems called analog-to-digital converters that sample and quantify analog signals to convert them into digital signals. After some processing of the digitized signal, a digital-to-analog converter can be used to convert the processed digital signal back into an analog signal. This analog electrical signal can then be converted into an acoustic signal though loudspeakers or earphones to make it available to human ears. These steps are represented in **Figure 1**.

Digital signal processing [1] is the set of theories and techniques for analyzing, synthesizing, quantifying, classifying, predicting, or recognizing signals, using digital systems.

A digital system receives as input a sequence of samples $\{x(0), x(1), x(2), \dots\}$, noted as $x(n)$, and produces as output a sequence of samples $y(n)$ after application of a series of algebraic operations.

A digital filter is a linear and invariant digital system. Let us consider a digital system that receives the sample sequences $x_1(n)$ and $x_2(n)$ as input. This system will respectively produce the sample sequences $y_1(n)$ and $y_2(n)$ as output. This system is linear if it produces the output $\alpha y_1(n) + \beta y_2(n)$ when it receives the sequence $\alpha x_1(n) + \beta x_2(n)$ as input. A digital system is said to be invariant if shifting the input sequence by n_0 samples also shifts the output sequence by n_0 samples.

These linear and invariant digital systems can be described by equations of the type:

$$\begin{aligned} y(n) + a_1 y(n-1) + a_2 y(n-2) + \dots + a_N y(n-N) \\ = b_0 x(n) + b_1 x(n-1) + \dots + b_M x(n-M) \end{aligned} \tag{1}$$

or

$$y(n) + \sum_{i=1}^{N} a_i y(n-i) = \sum_{i=0}^{M} b_i x(n-i) \tag{2}$$

This is equivalent to saying that the output $y(n)$ is a linear combination of the last N outputs, the input $x(n)$, and the M previous inputs. A digital filter is therefore determined if the coefficients a_i and b_i are known. A filter is called non-recursive if only the inputs are used to compute $y(n)$. If at least one of the previous output samples is used, it is called a recursive filter.

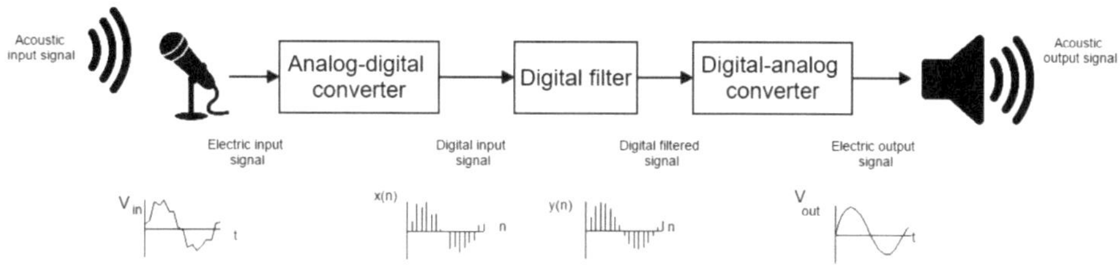

Figure 1.
Digital signal processing for acoustic signals.

2.2 Speech features

Speech is a signal carrying a lot of information. These expend from the sequence of words used to create a sentence, to the tone of voice used to utter this sentence. Not all of them are necessary to process and for some systems, trying to process all of them can harm the efficiency of the system. Also, the speech can carry noise before reception. That is why an important step in speech analysis is to extract descriptors or features that are relevant to the task of interest.

There exist many different feature spaces that describe speech information. In this section, we give an intuitive explanation of the ones widely used in Deep Learning architectures.

2.2.1 Power spectral density and spectrogram

Fourier analysis demonstrates that any physical signal can be decomposed into a sum of sinusoids of different frequencies. The power spectral density of a signal describes the amount of power carried by the different frequency bands of this signal.

This range of frequencies may be a discrete value set or a continuous frequency spectrum. In the field of digital signal processing, this power spectral density can be calculated by the Fast Fourier Transform (FFT) algorithm.

The graph of the power spectral density allows to visualize the frequency characteristics of a signal such as the fundamental frequency of a periodic signal and its harmonics. A periodic signal is a signal whose period is repeated indefinitely. The number of periods per unit of time that repeats is the fundamental frequency. Harmonics are the multiple frequencies of the fundamental. These frequencies have an important power density and present therefore extrema in the power spectral density.

An example of power spectral density is shown in the upper part of **Figure 2**. The first maximum is at the fundamental frequency which is 145.5 Hz. The other maxima are the harmonics.

When the signal's characteristics are evolving in the time, like with the voice for example, the spectrogram can be used to visualize this evolution. The spectrogram represents the power spectral density over the time. An example of power spectrogram is shown in the lower part of **Figure 2**. The x-axis is time and the y-axis is frequency. The colors correspond to the power density. A color scale is given on the right of the graph. The spectrogram is thus constructed by juxtaposing power spectral density functions computed on every frame as suggested in **Figure 2**.

2.2.2 Mel-spectrogram

The Mel-Spectrogram is a reduced version of the spectrogram. The use of this feature is very widespread for machine learning-based systems in general and for Deep learning-based TTS in particular.

The intuition behind this feature is to compress the representation of the speech in the higher values of the frequency domain based on the fact that human ear is sensitive to some frequencies more than others. The Mel Scale is an experimental function representing the sensitivity of human ear depending on the frequency.

The conversion of frequency f in Mel-frequency m is:

$$m = 2595 \cdot \log_{10}\left(1 + \frac{f}{700}\right) \quad (3)$$

Figure 3 shows the curve of the Mel Scale as a function of the frequency. As one can observe, an interval of low frequencies is mapped to a larger interval of Mel

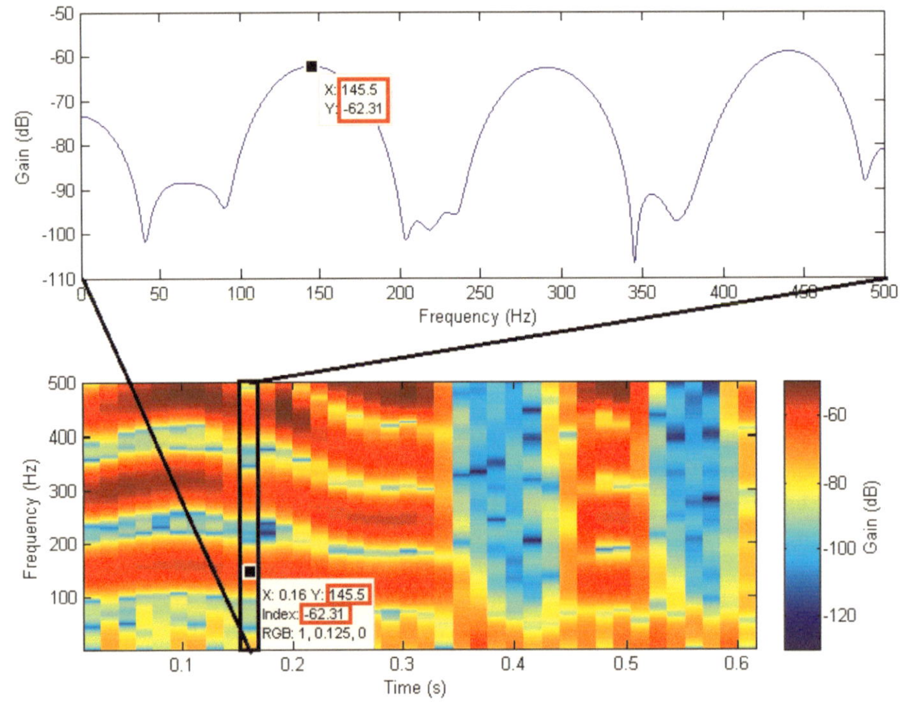

Figure 2.
Spectrum (top) and spectrogram (bottom) of a speech segment.

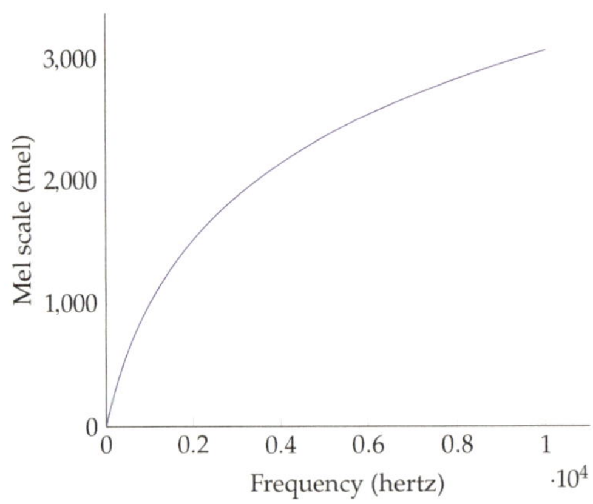

Figure 3.
Mel Scale representing the perception of frequencies.

values than for high frequencies. As an example, the interval $[0, 2000]$ Hz is mapped to more than 1500 Mel while the interval $[8000, 10000]$ Hz is mapped to less than 300 Mel.

3. Modeling of emotion expressiveness

Emotion modeling is one of the main challenges in developing more natural human-machine interfaces. Among the many existing approaches, two of them are widely used in applications. A first representation is Ekman's six basic emotion model [2], which identifies anger, disgust, fear, happiness, sadness, and surprise as six basic categories of emotions from which the other emotions may be derived.

Emotions can also be represented in a multidimensional continuous space like in the Russel's circumplex model [3]. This model makes it possible to better reflect the complexity and the variations in the expressions, unlike the category system. The two most commonly used dimensions in the literature are arousal, corresponding to the level of excitation, and valence, corresponding to the pleasantness level or positiveness of the emotion. A third dimension is sometimes added: dominance corresponding to the level of power of the speaker relative to the listener.

A more recent way of representing emotions is based on ranking, which prefers a relative preference method to annotate emotions rather than labeling them with absolute values [4]. The reason is that humans are not reliable for assigning absolute values to subjective concepts. However, they are better at discriminating between elements shown to them. Therefore, the design of perception tasks, for example, about emotion or style in speech, should take this into account by asking participants to solve comparison tasks rather than rating tasks.

It is important to note that many other approaches exist [5] and it is a difficult question to know what approach should be used in applications in the field of Human-Computer Interaction. Indeed, these psychological models of affect are propositions of explanations of how emotions are expressed. But these propositions are difficult to assess in practice.

Humans express their emotions via various channels: face, gesture, speech, etc. Different people will express and perceive emotions differently depending on their personality, their culture, and many other aspects. For developing application, one has therefore to take assumptions to reduce its scope and choose one approach of emotion modeling.

In this chapter, we are interested in how the expressive speech synthesized will be perceived. It is therefore reasonable to begin by choosing a language and assuming the origin of the synthesized voice.

Research has recently evolved into systems using, without preprocessing, the signal or spectrogram of the signal as input: the neural network learns the features that best correspond to the task it is supposed to perform on its own. This principle has been successfully applied to the modeling of emotions, currently constituting the state of the art in speech emotion recognition [6, 7].

4. Expressive speech synthesis

4.1 A brief history of speech synthesis techniques and how to control expressiveness

The goal behind a speech synthesis system is to generate an audio speech signal corresponding to any input text.

A sentence is constituted of characters and a human knows how these characters should be pronounced. If we want a machine to be able to generate speech signal from text, we have to teach it, or program it to do the same.

Such systems have been developed for decades and many different approaches were used. Here, we summarize them in three categories: Concatenation, Parametric Speech Synthesis, and Statistical Parametric Speech Synthesis. However, the state of the art is more diverse and complex. It contains many variants and hybrid approaches between them.

4.1.1 Concatenation

This approach is based on the concatenation of pieces of audio signals corresponding to different phonemes. This method is segmented in several steps. First, the characters should be converted in the corresponding phones to be pronounced. A simplistic approach is to assume that one letter corresponds to one phoneme for example. Then the computer must know what signal corresponds to a phoneme. A possibility to solve this problem is to record a database containing all the existing phonemes in a given language.

However concatenating phones one after another leads to very unnatural transitions between them. In the literature, this problem was tackled by recording successions of two phonemes, called diphones, instead of phones. All combinations of diphones are recorded in a dataset. The generation of speech is then performed by concatenation of these diphones. In this approach, many assumptions are not met in practice.

First, a text processing has to be performed. Indeed, text is constituted of punctuation, numbers, abbreviations, etc. Moreover, the letter-to-sound relationship is not respected in English and in many other languages. The pronunciation of words often depend on the context. Also, concatenating phone leads to a chopped signal and prosody of the generated signal is unnatural. To have a control on expressiveness with diphone concatenation techniques, it is possible to change $F0$ and duration with signal processing techniques implying some distorsion on the signal. Other parameters cannot be controlled without altering the signal leading to unnatural speech.

Another approach that is also based on the concatenation of pieces of signal is Unit Selection. Instead of concatenating phones (or diphones), larger parts of words are concatenated. An algorithm has to select the best units according to criteria: few discontinuities in the generated speech signal, a consistent prosody, etc.

For this purpose, a much larger dataset must be recorded containing a large variety of different combinations of phone series. The machine must know what part of signal corresponds to what phoneme, which means it has to be annotated by hand accurately. This annotation process is time-consuming. Today, there exist tools to do this task automatically. But this automation can in fact be done at the same time as synthesis as we will see later.

The advantages of this method is that the signal is less altered and most of the transitions between phones are natural because they are coming as is from the dataset.

With this method, a possibility to synthesize emotional speech is to record a dataset with separate categories of emotion. In synthesis, only units coming from a category will be used [8]. The drawback is that it is limited to discrete categories without any continuous control.

4.1.2 Parametric speech synthesis

Parametric Speech Synthesis is based on modeling how the signal is generated. It allows interpretability of the process. But in general, simplistic assumptions have to be made for speech modeling.

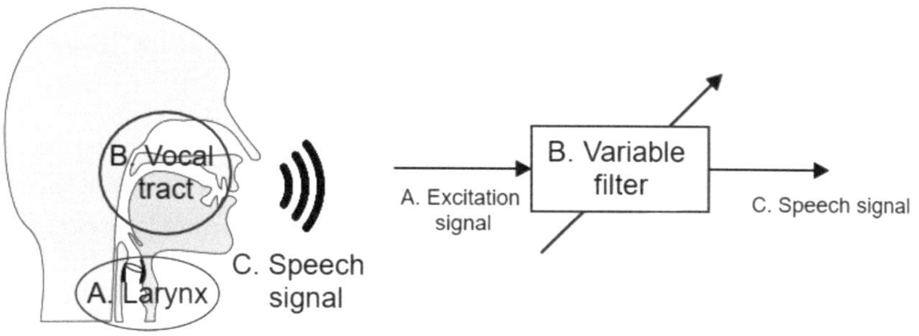

Figure 4.
Diagram describing voice production mechanism and source-filter model.

Anatomically, the speech signal is generated by an excitation signal generated in the larynx. This excitation signal is transformed by resonance through the vocal tract, which acts as a filter constituted by the guttural, oral, and nasal cavities. If this excitation signal is generated by glottal pulses, then a voiced sound is obtained. Glottal pulses are generated by a series of openings and closures of vocal cords or vocal folds. The vibration of the vocal chords has a fundamental frequency.

As opposed to voiced sounds, when the excitation signal is a simple flow of exhaled air, it is an unvoiced sound.

The source-filter model is a way to represent speech production, which uses the idea of separating the excitation and the resonance phenomenon in the vocal tract. It assumes that these two phenomena are completely decoupled. The source corresponds to the glottal excitation and the filter corresponds to the vocal tract. This principle is illustrated in **Figure 4**[1].

An example of Parametric Speech modeling is the linear prediction model. The linear prediction (LP) model uses this theory assuming that the speech is the output signal of a recursive digital filter, when an excitation is received at the input. In other words, it is assumed that each sample can be predicted by a linear combination of the last p samples. The linear predictive coding works by estimating the coefficients of this digital filter representing the vocal tract. The number of coefficients to represent the vocal tract has to be chosen. The more coefficients we take, the better the vocal tract is represented, but the more complex the analysis will be. The excitation signal can then be computed by applying the inverse filter on the speech signal.

In synthesis, this excitation signal is modeled by a train of impulses. In reality, the mechanics of the vocal folds is more complex making this assumption too simplistic.

The vocal tract is a variable filter. Depending on the shape we give to this vocal tract, we are able to produce different sounds. A filter is considered constant for a short period of time and a different filter has to be computed for each period of time.

This approach has been successful to synthesize intelligible speech but not natural human sounding speech.

For expressive speech synthesis, this technique has the advantage of giving access to many parameters of speech allowing a fine control.

The approach used in [9] to discover how to control a set of parameters to obtain a desired emotion was done through perception tests. A set of sentences were synthesized with different values of these parameters. These sentences were then

[1] Vocal tract image from: https://en.wikipedia.org/wiki/User:Tavin##/media/File:VocalTract.svg-Tavin/CC-BY-3.0

used in listening tests in which participants were asked to answer questions about the emotion they perceived. Based on these results, values of the different parameters were associated to the emotion expressions.

4.1.3 Statistical parametric speech synthesis

Statistical Parametric Speech Synthesis (SPSS) is less based on knowledge, and more based on data. It can be seen as Parametric Speech synthesis in which we take less simplistic assumptions on the speech generation and rely more on the statistics of data to explain how to generate speech from text.

The idea is to teach a machine the probability distributions of signal values depending on the text that is given. We generally assume that generating the values that are most likely is a good choice. We thus use the Maximum Likelihood principle (see Section 4.3.3).

These probability distributions are estimated based on a speech dataset. To be a good estimation of the reality, this dataset must be large enough.

The first successful SPSS systems were based on hidden Markov models (HMMs) and Gaussian Mixture models (GMMs).

The most recent statistical approach uses DNN [10], which is the basis of new speech synthesis systems such as WaveNet [11] and Tacotron [12]. The improvement provided by this technique [13] comes from the replacement of decision trees by DNNs and the replacement of state prediction (HMM) by frame prediction.

In the rest of this chapter, we focus on this approach of Speech Synthesis. Section 4.2 explains Deep Learning focusing in Speech Synthesis application and Section 4.3 reminds principles of Information Theory and probability distributions important in Speech Processing.

4.1.4 Summary

Depending on the synthesis technique used [14], the voice is more or less natural and the synthesis parameters are more or less numerous. These parameters allow to create variations in the voice. The number of parameters is therefore important for the synthesis of expressive speech.

While parametric speech synthesis can control many parameters, the resulting voice is unnatural. Synthesizers using the principle of concatenation of speech segments seem more natural but allow the control of few parameters.

The statistical approaches allow to obtain a natural synthesis as well as a control of many parameters [15].

4.2 Deep learning for speech synthesis

Machine Learning consists of teaching a machine to perform specific task, using data. In this chapter, the task we are interested in is Controllable Expressive Speech Synthesis.

The mathematical tools for this come from the field of Statistical Modeling.

Deep Learning is the optimization of a mathematical model, which is a parametric function with many parameters. This model is optimized or *trained* by comparing its predictions to ground truth examples taken from a dataset. This comparison is based on a measure of similarity or error between a prediction and the true example of the dataset. The goal is then to minimize the error or maximize the similarity. This can always be formulated as the minimization of a loss function.

To find a good loss function, it is necessary to understand the statistics of the data we want to predict and how to compare them. For this, concepts from information theory are used.

4.2.1 Different operations and architectures

The form of the mathematical function used to process the signal can be constituted of lots of different operations. Some of these operations were found very performant in different fields and are widely used. In this section, we describe some operations relevant for speech synthesis. In Deep Learning, the ensemble of the operations applied to a signal to have a prediction is called *Architecture*. There is an important research interest in designing architectures for different tasks and data to process. This research reports empirical results comparing the performance of different combinations. The progress of this field is directly related to the computation power available on the market.

Historically, the root of Deep Learning is a model called Neural Network. This model was inspired by the role of neurons in brain that communicate with electrical impulses and process information.

Since then, more recent models drove away from this analogy and evolves depending on their actual performance.

Fully connected neural networks are successions of linear projections followed by non-linearities (sigmoid, hyperbolic tangent, etc.) called layers.

$$h = f_h(W_h x + b_h) \tag{4}$$

x: input vector
h: hidden layer vector
W and b: parameter matrices and vector
f_h: Activation functions

More layers implies more parameters and thus a more complex model. It also means more intermediate representations and transformation steps. It was shown that deeper Neural Networks (more layers) performed better than shallow ones (fewer layers). This observation lead to the names Deep Neural Networks (DNNs) and Deep Learning. A complex model is capable of modeling a complex task but is also more costly to optimize in terms of computation power and data.

Merlin [16] toolkit has been an important tool to investigate the use of DNNs for speech synthesis. The first models developed within Merlin were based only on Fully connected neural networks. One DNN was used to predict acoustic parameters and another one to predict phone durations. It was a first successful attempt that outperformed other statistical approaches at the time.

Time dependencies are not well modeled and it ignores the autoregressive nature of speech signal. In reality, this approach relies a lot on data and does not use enough knowledge.

Convolutional Neural Networks (CNNs) refer to the operation of convolution and remind the convolution filters of signal processing (Eq. 5). A convolution layer can thus be seen as a convolutional filter for which the coefficients were obtained by training the Deep Learning architecture.

$$g(x, y) = \omega * f(x, y) = \sum_{s=-a}^{a} \sum_{t=-b}^{b} \omega(s,t) f(x-s, y-t) \tag{5}$$

f: input matrix
g: output matrix
ω: convolutional filter weights

Convolutional filters were studied in the field of image processing. We know what filters to apply to detect edges, to blurr an image, etc.

In practice, often, the operation implemented is correlation, which is the same operation except that the filter is not flipped. Given that the parameters of the filters are optimized during training, the flipping part is useless. We can just consider that the filter optimized with a correlation implementation is just the flipped version of the one that would have been computed if convolution was implemented.

For speech synthesis, convolutional layers have been used to extract a representation of linguistic features and predict spectral speech features.

For a temporal signal such as speech, one-dimensional convolution along the time axis allows to model time dependencies. As layers are stacked, the receptive field increases proportionally. In speech, there are long-term dependencies in the signal, for example, in the intonation and emphasis of some words. To model these long-term dependencies, dilated convolution was proposed. It allows to increase the receptive field exponentially instead of proportionally with the number of layers.

Recurrent Neural Network involves a recursive behaviour, that is, having an information feedback from the output to the input. This is analogous to recursive filters. Recursive filters are filters designed for temporal signals because they are able to model causal dependencies. It means that at a give time t, the value depends on the past values of the signal.

$$h_t = f_h(W_h x_t + U_h h_{t-1} + b_h) \tag{6}$$

$$y_t = f_y(W_y h_t + b_y) \tag{7}$$

x_t: input vector.
h_t: hidden layer vector
y_t: output vector
W, U and b: parameter matrices and vector
f_h and f_y: activation functions

4.2.2 Encoder and decoder

An encoder is a part of neural network that outputs a hidden representation (or latent representation) from an input. A decoder is a part of neural network that retrieves an output from a latent representation.

When the input and the output are the same, we talk about auto-encoders. The task in itself is useless, but the interesting part here is the latent representation. The latent space of an auto-encoder can provide interesting properties such as a lower dimensionality, meaning a compressed representation of the initial data or meaningful distances between examples.

4.2.3 Sequence-to-sequence modeling and attention mechanism

A sequence-to-sequence task is about converting sequential data from one domain to another, for example, from a language to another (translation), from speech to text (speech recognition), or from text to speech (speech synthesis).

First Deep Learning architectures for solving sequence-to-sequence tasks were based on encoder-decoders with RNNs called RNN transducer.

Other techniques were were found to outperform this. The use of Attention Mechanism was found beneficial [17].

Attention Mechanism was first developed in the field of computer vision. It was then successfully applied to Automatic Speech Recognition (ASR) and then to Text-to-Speech synthesis (TTS).

In the Deep Learning architecture, a matrix is computed and used as weighting on the hidden representation at a given layer. The weighted representation is fed to the rest of the architecture until the end. This means that the matrix is asked to emphasize the part of the signal that is important to reduce the loss. This matrix is called the Attention matrix because it represents the importance of the different regions of the data.

In computer vision, a good illustration of this mechanism is that for a task of classification of objects, the attention matrix has high weights for the region corresponding to the object and low weights corresponding to the background of the image.

In ASR, this mechanism has been used in a so-called *Listen, Attend and Spell* [18] (LAS) setup. An important difference compared to the previous case is that it is a sequential problem. There must be an information feedback to have a recursive kind of architecture and each time step must be computed based on previous time steps.

LAS designates three parts of the Deep Learning architecture. The first one encodes audio features in a hidden representation. The role of the last one is to generate text information from a hidden representation. Between this encoder and decoder, at each time step, an Attention Mechanism computes a vector that will weigh the text encoding vector. This weighting vector should give importance to the part of the utterance to which the architecture should pay attention to generate the corresponding part of speech.

An Attention plot (**Figure 5**) of a generated sentence can be constructed by juxtaposing all the weighting vectors computed during the generation of a sentence. The resulting matrix can then be represented by mapping a color scale on the values contained.

This attention plot shows an attention path, that is, the importance given to characters along the audio output timeline. As can be observed in **Figure 5**, this attention path should have a close to diagonal shape. Indeed, the two sequences have a close chronological relationship.

4.3 Information theory and speech probability distributions

Information Theory is about optimizing how to send messages with as few resources as possible. To that end, the goal is to compress the information by

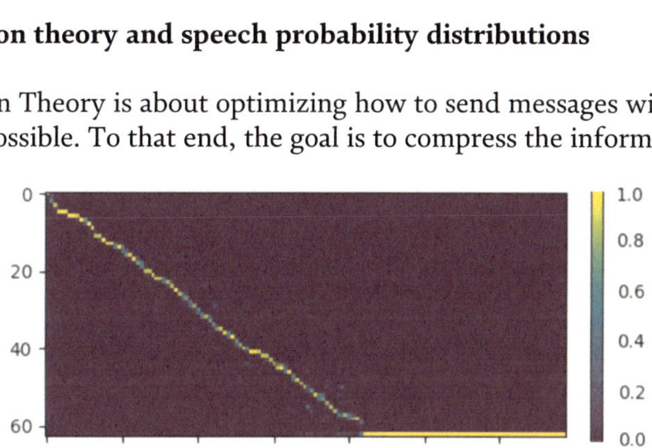

Figure 5.
Alignment plot. The y-axis represents the character indices and the x-axis represents the audio frame feature indices. The color scale corresponds to the weight given to a given character to predict a given audio frame.

using the right code so that the messages do not contain redundancies to be as small as possible.

4.3.1 Information and probabilities

Shannon's Information Theory quantifies information, thanks to the probability of outcomes. If we know an event will occur, its occurrence gives no information. The less likely it is to happen, the more it gives information.

This relationship between information and probability of an event is given by Shannon information content measured in bits. A bit is a variable that can have two different values: 0 or 1.

$$h(x) = \log_2\left(\frac{1}{p(x)}\right) \tag{8}$$

The number of possible messages with L bits is 2^L. If all messages are equally probable, the probability of each message is $p = \frac{1}{2^L}$. We then have $L = \log_2\left(\frac{1}{p}\right)$. A generalization of this formula in which the messages are not equally probable is Eq. (8). It can be interpreted as the minimal number of bits to communicate this message.

The probability represents the degree of belief that an event will happen [19]. For example, we can wonder the probability of a result of four by rolling a six-sided die or the probability that the next letter in a text is the letter *r*.

These probabilities depend on the assumptions we make:

- Is the die perfectly balanced? If yes, the probability of a result of four is 1/6.
- What is the language of the text? Do we know the subject, etc. Depending on this information, we can have different estimations of this probability.

We obtain a probability distribution by listing the probability of all the possible outcomes. For the example of the result by rolling the perfectly balanced die, the possible outcomes are $[1, 2, 3, 4, 5, 6]$ and their probabilities are $[1/6, 1/6, 1/6, 1/6, 1/6, 1/6]$.

In both examples, we have a finite number of possible outcomes. The probability distribution is said to be discrete. On the contrary, when the possible outcomes are distributed on a continuous interval, then the probability distribution is said to be continuous. This is the case, for example, of amplitude values in a spectrogram.

The most famous continuous probability distribution is the Gaussian distribution:

$$p(x) = \frac{1}{\sigma\sqrt{2\pi}} e^{-\frac{1}{2}\left(\frac{x-\mu}{\sigma}\right)^2} \tag{9}$$

Another important distribution, especially in speech processing, is the Laplacian distribution:

$$p(x) = \frac{1}{2b} e^{\left(-\frac{|x-\mu|}{b}\right)} \tag{10}$$

Both distributions are plotted in **Figure 6**. The blue curve corresponds to the Gaussian probability distribution (with $\mu = 0$ and $\sigma = 0.5$) and the red curve corresponds to the Laplacian probability distribution (with $\mu = 0$ and $b = 0.5$). For both distributions, the maximum is μ, and are symmetrically decreasing as the distance from μ increases.

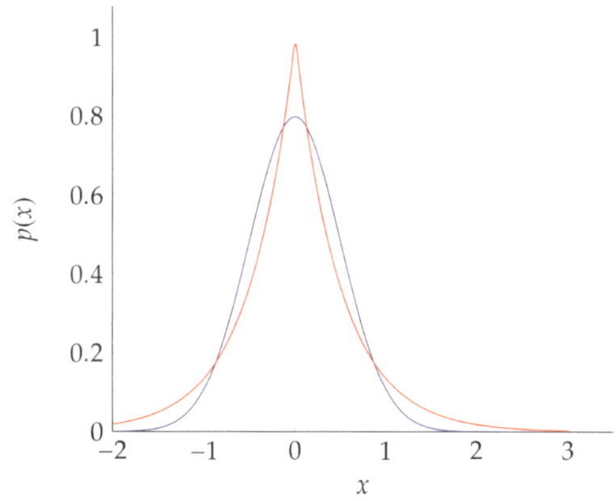

Figure 6.
In blue: Gaussian distribution with $\mu = 0$ *and* $\sigma = 0.5$*. In red: Laplacian distribution with* $\mu = 0$ *and* $\mathrm{b} = 0.5$*.*

4.3.2 Entropy and relative-entropy

The average information content of an outcome, also called entropy, of the probability distribution p is:

$$H(p) = \sum_x p(x) \log_2 \left(\frac{1}{p(x)} \right) \tag{11}$$

The relative-entropy between two probability distributions, also called Kullback-Leibler divergence, is defined as:

$$D_{\mathrm{KL}}(p \| q) = \sum_x p(x) \log \frac{p(x)}{q(x)} \; dx \tag{12}$$

It represents a dissimilarity between two probability distributions.

4.3.3 Maximum likelihood and particular cases

This concept is necessary to understand how to train a Deep Learning algorithm or, more generally, how to find the optimal parameters of a model. The role of a statistical model is to represent as accurately as possible the behavior of a probability distribution.

Maximum likelihood estimation (MLE) (Eq. 13) allows to estimate the parameters θ of a statistical parametric model $p(x|\theta)$ by maximizing the probability of a dataset under the assumed statistical model, that is, the Deep Learning architecture.

$$\theta_{MLE} = \arg\max p(\mathbf{x}|\theta) \tag{13}$$

It can be demonstrated that this is equivalent to minimizing $D_{\mathrm{KL}}(p \| q)$ with p, the probability distribution of the model and q, the probability of the data [20]. It is a way to express that the probability distribution generated by the model should be as close as possible to the probability distribution of the data.

If assumptions can be made on the probability distributions, it is possible to have distances or errors for which the minimization is equivalent to MLE. These errors are computed by comparing estimations from the model $\hat{Y}_i$ and the value from the dataset Y_i.

Maximizing likelihood assuming a Gaussian distribution is equivalent to minimizing Mean Squared Error (MSE):

$$MSE = \frac{1}{n}\sum_{i=1}^{n}\left(Y_i - \hat{Y}_i\right)^2 \tag{14}$$

Maximizing likelihood assuming a Laplacian distribution is equivalent to minimizing Mean Absolute Error (MAE):

$$MAE = \frac{1}{n}\sum_{i=1}^{n}\left|Y_i - \hat{Y}_i\right| \tag{15}$$

To choose the right criterion to optimize when working with speech data, one should pay attention to speech probability distributions. Speech waveforms and magnitude spectrogram distribution are Laplacians [21, 22]. That is why MAE loss should be used to optimize their predictions.

5. Summary and application

In this chapter, we first briefly introduced digital signal processing and digital filtering, and described the different possibilities of emotion representation and the few most important speech feature spaces in this context, namely spectrogram and Mel-spectrogram.

Available speech synthesis methods were then exposed, from concatenation of speech signal segments to parametric modeling of speech production, to statistical parametric speech synthesis.

Most recent SPSS systems use Deep Learning that can be seen as non-linear signal processing for which filters are optimized based on data.

We focused on the tools for SPSS and explained Deep Learning architecture blocks that are used along with the right loss functions based on the probability distributions of speech features.

To build a controllable expressive speech synthesis system, one should keep several concepts in mind. First, it is necessary to gather data and process them to have a good representation to be used with a Deep Learning algorithm, that is, text, Mel-spectrograms, and information about the expressivness of speech. Then one has to design a Deep Learning architecture. Its operations should be inspired by the features to model (1D convolution or RNN cells for long-term context, attention mechanism for recursive relationships). It should have a way to control expressiveness either with a categorical representation [23] or a continuous representation [24]. But it is important to take into account that annotations should not be acquired from humans by asking them to give absolute values on subjective concepts, but rather by asking them to compare examples. And finally, the parametric model should be trained with a loss function adapted to the probability distribution of the acoustic features, that is, MAE and Kullback-Leibler divergence loss.

Acknowledgements

Noé Tits is funded through a PhD grant from the Fonds pour la Formation à la Recherche dans l'Industrie et l'Agriculture (FRIA), Belgium.

Author details

Noé Tits*, Kevin El Haddad and Thierry Dutoit
Numediart Institute, University of Mons, Mons, Belgium

*Address all correspondence to: noe.tits@umons.ac.be

References

[1] Dutoit T, Marques F. Applied Signal Processing: A MATLABTM-based proof of concept. Springer Science & Business Media; 2010

[2] Ekman P. An argument for basic emotions. Cognition & Emotion. 1992;**6** (3–4):169-200

[3] Russell JA. A circumplex model of affect. Journal of Personality and Social Psychology. 1980;**39**(6):1161

[4] Yannakakis GN, Cowie R, Busso C. The ordinal nature of emotions. In: 2017 Seventh International Conference on Affective Computing and Intelligent Interaction (ACII). 2017. pp. 248-255. Available from: doi.ieeecomputersociety.org/10.1109/ACII.2017.8273608

[5] Barrett LF. The theory of constructed emotion: An active inference account of interoception and categorization. Social Cognitive and Affective Neuroscience. 2017;**12**(1):1-23

[6] Martinez HP, Bengio Y, Yannakakis GN. Learning deep physiological models of affect. IEEE Computational Intelligence Magazine. 2013;**8**(2):20-33

[7] Trigeorgis G, Ringeval F, Brueckner R, Marchi E, Nicolaou MA, Schuller B, et al. Adieu features? End-to-end speech emotion recognition using a deep convolutional recurrent network. In: 2016 IEEE international conference on Acoustics, Speech and Signal Processing (ICASSP). 2016. pp. 5200-5204

[8] Schröder M. Emotional speech synthesis: A review. In: Seventh European Conference on Speech Communication and Technology. 2001

[9] Burkhardt F, Sendlmeier WF. Verification of acoustical correlates of emotional speech using formant-synthesis. In: ISCA Tutorial and Research Workshop (ITRW) on Speech and Emotion. 2000

[10] Zen H, Senior A, Schuster M. Statistical parametric speech synthesis using deep neural networks. In: 2013 IEEE International Conference on Acoustics, Speech and Signal Processing (ICASSP). 2013. pp. 7962-7966

[11] van den Oord A, Dieleman S, Zen H, Simonyan K, Vinyals O, Graves A, et al. WaveNet: A generative model for raw audio. In: SSW. arXiv preprint arXiv: 1609.03499. 2016

[12] Wang Y, Skerry-Ryan RJ, Stanton D, Wu Y, Weiss RJ, Jaitly N, et al. Tacotron: Towards end-to-end speech synthesis. In: INTERSPEECH. 2017

[13] Watts O, Henter GE, Merritt T, Wu Z, King S. From HMMs to DNNs: Where do the improvements come from? In: 2016 IEEE International Conference on Acoustics, Speech and Signal Processing (ICASSP). 2016. pp. 5505-5509

[14] Burkhardt F, Campbell N. Emotional speech synthesis. In: The Oxford Handbook of Affective Computing. New York: Oxford University Press; 2014. 286p

[15] Zen H, Tokuda K, Black AW. Statistical parametric speech synthesis. Speech Communication. 2009;**51**(11): 1039-1064

[16] Wu Z, Watts O, King S. Merlin: An open source neural network speech synthesis system. In: Proc SSW, Sunnyvale, USA. 2016

[17] Prabhavalkar R, Rao K, Sainath TN, Li B, Johnson L, Jaitly N. A comparison of sequence-to-sequence models for

speech recognition. In: INTERSPEECH. 2017. pp. 939-943

[18] Chan W, Jaitly N, Le Q, Vinyals O. Listen, attend and spell: A neural network for large vocabulary conversational speech recognition. In: 2016 IEEE International Conference on Acoustics, Speech and Signal Processing (ICASSP). 2016. pp. 4960-4964

[19] MacKay DJ, Mac Kay DJ. Information Theory, Inference and Learning Algorithms. Cambridge, UK: Cambridge University Press; 2003. Available from: https://www.cambridge.org

[20] Goodfellow I, Bengio Y, Courville A. Deep Learning. Cambridge, Massachusetts: MIT Press; 2016. Available from: https://mitpress.mit.edu/about

[21] Gazor S, Zhang W. Speech probability distribution. IEEE Signal Processing Letters. 2003;**10**(7):204-207

[22] Usman M, Zubair M, Shiblee M, Rodrigues P, Jaffar S. Probabilistic modeling of speech in spectral domain using maximum likelihood estimation. Symmetry. 2018;**10**(12):750

[23] Tits N, Haddad K, Dutoit T. Exploring transfer learning for low resource emotional TTS. In: Bi Y, Bhatia R, Kapoor S, editors. Intelligent Systems and Applications. Cham: Springer International Publishing; 2020. pp. 52-60

[24] Tits N, Wang F, Haddad KE, Pagel V, Dutoit T. Visualization and interpretation of latent spaces for controlling expressive speech synthesis through audio analysis. In: Proceedings of INTERSPEECH. 2019. pp. 4475-4479. Available from: http://dx.doi.org/10.21437/Interspeech.2019-1426

Chapter 4

Speech Standards: Lessons Learnt

Paolo Baggia

Abstract

During the past decades, the landscape of speech and DTMF applications has changed from being based on proprietary platforms to being completely based on speech standards. The W3C Voice Browser Working Group played a primary goal in this change. This chapter describes that change, highlights the standards created by the W3C VBWG, and discusses the benefits that these standards have provided in many other application fields, including multi-modal interfaces.

Keywords: human-computer interaction, speech recognition, speech synthesis, pronunciation lexicons, natural language, spoken dialog, speech grammars, multi-modal interaction

1. Introduction

This chapter is a retrospective of a very special moment for speech and dialog technologies. Since the end of the last century, the use of speech and dialog technologies has been limited by proprietary implementation of platforms with strong legacies, as well as limitations on the rapid adoption of the core technology advances. But at the turn of the century, a sudden change occurred, and speech applications started to be deployed in many commercial applications, becoming ubiquitous in a matter of years. That trend has continued to this day. This transformation was catalyzed by the creation and adoption of a family of standard languages. This evolution was quickly accepted and adopted by the industry, even before these languages were completely defined. In the meantime, research was constantly increasing performance, which also fueled the widespread deployment of speech applications.

In the early 2000s, a sharp increase in the number of speech applications occurred in many areas, including customer care, finance, travel, and many other sectors. This increase is continuing, with the diffusion of virtual assistants, automatic chatbots, speech presence in smartphones, in the car, and at home. Speaking to an appliance is now an everyday activity, while it was a dream limited to sci-fi movies only a few decades earlier. That dream is a reality today and industrial standards have played a significant role in this achievement.

The ecosystem created by speech standards has been active for more than 20 years, and it is in the core of the major players, even though there are recent trends to move from in-house technologies to hosting and to access speech resources by Web APIs. In these new developments, speech standards can still play a role to provide customizations to hosted resources.

IntechOpen

The success of this enterprise was made possible by a highly collaborative work among a large group of people, from academia to industries and even individual contributors. The hope is that this example will inspire new developments in the future, and research and industry will be ready to create a new open ecosystem.

2. Why and when?

At the beginning of this century, the time was ready for a change of paradigm in the way speech technologies were deployed.

In the previous decade, research had been constantly improving the accuracy and powerfulness of speech technologies. For instance:

- Automatic speech recognition (ASR) moved from very limited tasks, such as digit recognition, to large vocabulary continuous speech recognition (LVCSR) by the adoption of statistical models (dynamic programming, hidden Markov models, statistical language models, etc.) The accuracy improvement was accelerated by government-sponsored competitions among the leading research labs and companies. These included DARPA funded projects such as the Airline Travel Information System (ATIS) [1–3], a speech understanding challenge focused on data collection of spoken flight requests, and the Wall Street Journal Continuous Speech Recognition Corpus [4], attempting to recognize speech from read WSJ articles.

- Speech synthesis and text-to-speech (TTS) during the 1980s reached the goal of high intelligibility and flexibility with a parametric approach [5], but the automatic voices were still robotic. A new technique, Concatenative Unit Selection [6], was less flexible, but capable of a more natural rendering and it was generally adopted by the industry.

- Spoken dialog systems (SDS) research was initially promoted by EU-funded projects, such as SUNDIAL [7], RAILTEL [8], and its continuation ARISE [9]. The results achieved in those projects were very promising to the point that the Italian Railways company (Ferrovie dello Stato, now Trenitalia) decided to deploy the prototype developed within the ARISE project with the help of Telecom Italia Labs (TILAB). The resulting phone service, known as FS_Informa, enabled customers to request train timetables over the phone. For a review of the state-of-the-art on Human Language Technology at that time, see [10], while for a comprehensive and accessible view of speech technologies, see [11].

Speech technologies were ready for commercial deployments, but there were many obstacles along the way. One major obstacle was that each technology company had its own proprietary APIs, to be integrated in a proprietary IVR platform. This slowed down the delivery of the latest technology advances because of the platform provider resistance to changing their proprietary environments. Also, customers were locked in on individual vendor's proprietary legacies.

Another important factor was the contemporaneous evolution of the Web infrastructure spearheaded by the W3C Consortium, led by Tim Berners-Lee, the Web's inventor. W3C, the World Wide Web Consortium is an international community whose mission is to drive the Web to its full potential by developing protocols and guidelines that ensure its long-term growth. This inspired researchers to consider whether a Web-based architecture could accelerate the evolution of speech applications. This was the idea behind a seminal event, a W3C Workshop held in

Cambridge (MA) on October 13, 1998 [12], promoted by Dave Raggett of the W3C and Dr. James A. Larson of Intel. The workshop was named: "Voice Browsers," as an event to discuss different innovative ideas on how to solve the proprietary issues by adopting the latest advances offered by the Web infrastructure. The workshop catalyzed the interest of research labs, companies, and start-ups, and it culminated in the creation of the W3C Voice Browser Working Group (VBWG) [13] chaired by Jim Larson and Scott McGlashan of PipeBeach (later Hewlett-Packard). Inside the W3C VBWG, a subgroup was devoted to study the expansion of the ideas in a multi-modal environment, and after a few years, it spun off a second group: the W3C Multi-Modal Interaction Working Group (MMIWG) [14], chaired by Dr. Deborah Dahl of Unisys (later Conversational Technologies).

The goal of the VBWG was to create a family of interoperating standards, while the MMIWG had the role to re-use those new standards in multi-modal applications, where other modalities were active in addition to voice for input and output (visual, haptic, etc.).

Figure 1 shows the initial diagram proposed by Jim Larson and named: "Speech Interaction Framework" (see the original diagram in Section 4 of [15], see also [16]). It remained the reference point for the development of all the standard languages created along the years.

The solid boxes are the modules of a reference spoken dialog architecture centered around the dialog manager, which is connected with an external telephony system and the Web. This shows the attempt to align the Web along with the main communication channel of the time. In this framework, there are input modules such as the ASR (automatic speech recognition) engine and a touch-tone (DTMF) recognizer. Additional modules include language understanding and context interpretation, but they were not considered to be priorities at that time. TTS (text-to-speech) engine, pre-recorded audio player, language generation, and media planning are considered output modules. After a considerable work by the W3C VBWG, the modules colored in red became completely driven by W3C Recommendations (the dashed red bordered boxes).

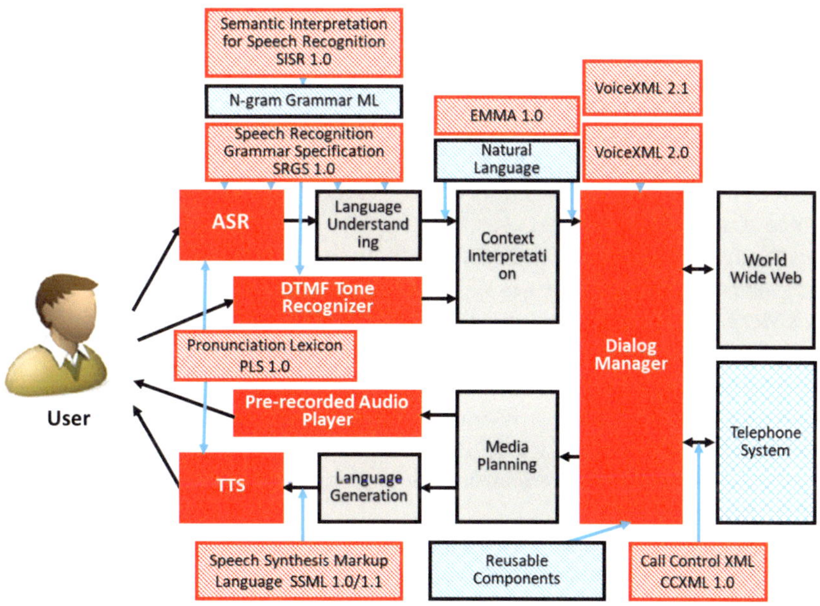

Figure 1.
Speech interaction framework.

From its creation, W3C VBWG started to attract all the companies and labs active in that space. The companies included speech technology providers (at that time L&H, Philips, Nuance, SpeechWorks, Loquendo, and Entropic), research labs (MIT, Rutgers, AT&T Bell Labs, and CSELT/TILAB), large telcos (Lucent, AT&T, BT, Deutsche Telekom, France Telecom, Telecom Italia, Motorola, and Nokia), large players (Microsoft, HP, Intel, IBM, and Unisys), and IVR vendors (Avaya, Genesys, Comverse, and CISCO). In addition, newly created companies such as voice platform providers (PipeBeach, Voxpilot, Vocalocity, VoiceGenie, and Voxeo), voice application host (HeyAnita, BeVocal, and Tellme), and many more joined the effort.

One of the first actions of the W3C VBWG was to acknowledge the contribution of the VoiceXML Forum [17] (founded by AT&T, Lucent, Motorola, and IBM) of a new markup language called VoiceXML 1.0 [18] of their design. From this point on, the W3C VBWG focused on completing VoiceXML with additional features. However, a wise decision was made to create a family of interoperable standards instead of a monolithic language. These standard languages are those described in Section 3. At the same time, the VoiceXML Forum took on a complementary role in the evolution of the VoiceXML ecosystem. It focused on education, evangelization, and support of the adoption of this family of standards. Among the major achievements of the VoiceXML Forum are the following two programs:

- The Platform Certification Program to allow platform developers to thoroughly test and certify that their platforms support all standard features. The first certification program was limited to VoiceXML 2.0 with a large adoption of 26 platforms certified. It was then extended to also certify VoiceXML 2.1, SRGS 1.0, and SSML 1.0 with eight more platforms certified.

- The Developers Certification Program to allow developers to certify their competence in the VoiceXML architecture and in the correlated standards.

All the materials produced are still available in the VoiceXML Forum Web site [17].

3. W3C VBWG standards

The W3C VBWG, supported by the VoiceXML Forum, accelerated a cooperative effort to create the foundations of a new generation of voice applications based on public standards. In a short time, an incredible sequence of Working Drafts was published, demonstrating the energy and creativity underlying the development of the voice standards.

In March 2004, after less than 4 years from the start of VBWG, the first group of complete standards, known as W3C Recommendations, was released. It includes VoiceXML 2.0 [19] for authoring voice applications; SRGS 1.0 [20] for defining the syntax of speech grammars; and SSML 1.0 [21] for controlling speech synthesis (or text-to-speech, TTS). A few years later, in April/June 2007, a second round of W3C Recommendations was released, which includes VoiceXML 2.1 [22], which completes VoiceXML 2.0 with a limited number of new features; and SISR 1.0 [23], which standardizes the creation of a meaning representation from a SRGS 1.0 speech grammar.

The work continued in the following years. SSML 1.0 was revised to version 1.1 [24] to improve the internationalization of speech synthesis in other regions of the world, including India and Eastern Asia, and PLS 1.0 [25], which supports the description of pronunciation lexicons, a shared resource for both SRGS 1.0 and SSML 1.0/1.1 resources. Finally, CCXML 1.0 [26] was released as a real-time language to implement telephony and VoIP call control in a voice browser platform, while SCXML 1.0 [27] as a general-purpose event-based state machine language that can be used for defining the dialog manager, and other components of a speech system. A comprehensive introduction to SCXML 1.0 is available in [28]. In the rest of this section, these languages will be briefly introduced.

3.1 Dialog management: VoiceXML 2.0/2.1

The Voice Extensible Markup Language (VoiceXML), version 2.0 [19], standard was the center of the innovation. Its key features are as follows:

- It is an XML declarative language.
- It is easy to author, the motto was: "Simple things must be easy and complex things must be possible!"[1]
- It assumes the existence of the Web architecture.

All these features carry clear advantages. An XML language allows a clean syntax checked by DTD/Schema, extensibility by namespaces, and encodings, generally available with any XML processors (user agent). The second feature, simplicity and flexibility, allows to edit VoiceXML 2.0 as text editor then upload it as a static page or generate it dynamically by Web applications (like all of the Web sites today). Finally, to be within the Web architecture means to share an enormous background of tools and techniques and it is part of the mainstream of the current technology evolution.

From a functional point of view, VoiceXML 2.0 allows the creation of speech applications that can replace menu-based, DTMF, and pre-recorded messages by a voice-driven interaction where the messaging is synthesized speech. This was the main reason why all the major IVR platforms quickly adopted VoiceXML, enabling taking advantage of a more powerful application environment. The second reason was the need to open the world of IVR applications to new players, instead of relying on proprietary solutions. Not only does VoiceXML 2.0 allow platforms to take advantage of the latest advances in ASR and TTS engines but also allow them to continue implementing traditional menu-based DTFM applications. Consequently, with VoiceXML, a complete replacement of the previous generation of IVRs became possible. More recently, VoiceXML 2.1 [22] further extended the language with additional features mostly devoted to creating more dynamic applications. This was a general trend in the evolution of the Web, as well as in the evolution in VoiceXML.

Figure 2 shows a simplified VoiceXML 2.0 document that implements a dialog to request departure and arrival airports from a user. The dialog tries first to recognize both the locations in a single utterance; if that fails, it asks them again in sequence. A final confirmation is given before transitioning to another page of the application. This is called a mixed-initiative dialog, where a user has a certain degree of freedom in expressing requests. For a detailed introduction of VoiceXML, see [29–31].

3.2 Speech recognition: SRGS 1.0 and SISR 1.0

Two standards were created by the W3C VBWG to define the knowledge resources for ASR engine: speech grammars and semantic interpretation. The first one is the formal definition of a speech grammar described in the W3C Recommendation "Speech

[1] The original quote is from Alan Key.

```
<?xml version="1.0" encoding="UTF-8"?>
<vxml version="2.0" xmlns="http://www.w3.org/2001/vxml"
  xml:lang="en-GB">
<form id="dep_arr_airports">
 <grammar src="dep_arr.grxml"
         type="application/srgs+xml"/>
 <initial name="start">
   <prompt>
         What are the arrival and departure airports?
   </prompt>
 </initial>
 <field name="fromcity">
   <prompt>Tell me the departure airport.</prompt>
 </field>
 <field name="tocity">
   <prompt>Tell me the arrival airport.</prompt>
 </field>
 <field name="go_ahead" type="boolean" modal="true">
   <prompt>Do you want to leave from
     <value expr="fromcity"/> and arrive
     to <value expr="tocity"/>?
   </prompt>
   <filled>
     <if cond="go_ahead">
       <submit next="/servlet/dep_date"
              namelist="fromcity tocity"/>
     </if>
     <clear namelist="fromcity tocity go_ahead"/>
   </filled>
 </field>
</form>
</vxml>
```

Figure 2.
A simplified VoiceXML document.

Recognition Grammar Specification Version 1.0" SRGS 1.0 [20]. Speech grammars and statistical language models (SLMs) are the two common ways to provide constraints to the speech recognition process. A grammar is a formal definition of all the sentences that can be spoken. The grammar drives the ASR engine to find the closest match with the acoustic signal. A grammar is a strong constraint for the ASR and is relatively simple to implement. Statistical LMs, typically used in speech-to-text systems where the user in not specifically prompted, are in contrast weaker constraints characterized by the probability of a word to be spoken in the context of the preceding words (known as n-gram probabilities). The W3C VBWG standardization effort focused on the speech grammar only, because it was useful for simpler recognition tasks, but also because the other formats, driven by research, were commonly used for n-grams[2]. A proposal for an SLM standard in the VBWG is described in [32].

SRGS supports the definition of grammars for speech as well as for DTMF inputs. A grammar can be specified in two equivalent formats, an XML document, called GrXML and a more traditional textual format, called ABNF, the acronym for augmented Backus-Naur format (commonly used to describe the syntax of a programming language). The W3C SRGS 1.0 Recommendation very clearly defines those two equivalent formats and offers a great number of examples (see [20]).

2 For instance, the well-known MIT ARPA LM format, see http://www.seas.ucla.edu/spapl/weichu/htkbook/node243_mn.html

The SRGS 1.0 specification was immediately adopted by all speech recognition engines, allowing them to interoperate within a VoiceXML platform. Of the two formats, GrXML became the predominant one, but it is very easy to transform a grammar from one format to the other.

Figure 3 shows an excerpt of a SRGS 1.0 grammar, in GrXML format, with the goal to recognize utterances like: "from Rome to Paris," where the list of cities might be extended to a longer list.

The part of the grammar devoted to the generation of a meaning representation or semantic interpretation is indicated with blue characters. This is the domain of the second speech grammar standard produced by W3C VBWG "Semantic Interpretation for Speech Recognition Specification Version 1.0" SISR 1.0 [23].

Semantic results are encapsulated in each rule by means of <tag> elements, which contain snippets of the programming language ECMAScript [33], widely known in its Web variety as JavaScript. The W3C SISR 1.0 Recommendation prescribes the use of the Compact Profile ECMA-327, which is a constrained version of ECMAScript. The goal was to gain computational efficiency to enable more compact speech recognition engine processing.

In SISR 1.0, each SRGS 1.0 rule, like "city" in **Figure 3**, contains a predefined variable called "out" whose properties are assigned within the <tag> elements. The content of the "out" variable of the most external rule, called the "root" rule, is returned from the recognition engine to the application environment.

For the input utterance "from Rome to Paris," for example, the SRGS grammar in **Figure 3** will return the ECMAScript object:

{fromcity: "FCO", tocity: "CDG"}

This is the case for simple and focused grammars where the result is just one or a few values. However, SISR supports also conditional logic and algorithms. This would be useful for instance to validate a checksum in a complex numeric (i.e., credit card numbers) or alphanumeric strings (as the personal taxation ID in Italy). That would allow the recognizer to validate and possibly reject a wrong result before returning it to the application and at the same time to increase the confidence of alternative, and possibly correct, recognition result.

```
<?xml version="1.0" encoding="UTF-8"?>
<grammar version="1.0" xml:lang="en-GB"
  xmlns="http://www.w3.org/2001/06/grammar"
  tag-format="semantics/1.0" root="fromto">

<rule id="fromto" scope="public">
   from <ruleref uri="#city"/>
   <tag>out.fromcity=rules.latest();</tag>
    to <ruleref uri="#city"/>
   <tag>out.tocity= rules.latest();</tag>
</rule>

<rule id="city">
     <one-of>
      <item>London<tag>out="LHR"</tag></item>
      <item>Paris<tag>out="CDG"</tag></item>
      <item>Rome<tag>out="FCO"</tag></item>
     </one-of>
</rule>
</grammar>
```

Figure 3.
Simple SRGS grammar with SISR script.

3.3 Speech synthesis: SSML 1.0 and 1.1

Another effort was to define how to control a speech synthesis, or TTS engine. This is to help the engine render the textual prompt in the most accurate way. The XML markup language for this purpose is the Speech Synthesis Markup Language Version 1.0, SSML 1.0 [20], which was released in March 2004.

Figure 4 shows the five major processing steps present in all TTS engines. For each of them, the engine offers a normal behavior, called "non-markup behavior" in the picture. The SSML mark-up instead allows the engine to improve the default rendering by means of elements of the language. Each element is related to one specific processing step, and it is interpreted as a request by the author to perform some specific processing. It is then up to the processor to determine whether and in what way to realize the command.

The SSML example in **Figure 5** shows a prompt for a flight information system structured into a single paragraph (<p > element) and two sentences (<s > elements). Acronyms are substituted (<sub>) into expanded versions, pauses are added (<break>), and a time expression is explicitly labeled (<say-as>) to select the correct way of reading it. Other elements can change additional features, such as prosodic features of speed and rate (<prosody>), and how to change the speaking voice (<voice>).

SSML 1.0 [21] continued to be standardized to promote the use of SSML to more international languages, in particular Asian and Indian languages. Three workshops were held to encourage local companies and universities to propose features to be added to the language:

- Nov 2005 at Beijing (China)
- May 2006 at Crete (Greece)
- Jun 2007 at Hyderabad

A new standard SSML 1.1 [24] was released in September 2010. See Appendix F of [24] for details on the changes. Among them, a < token> element was introduced for languages where the whitespace has peculiar behavior, such as in Mandarin, Japanese, Thai, Vietnamese, and Urdu.

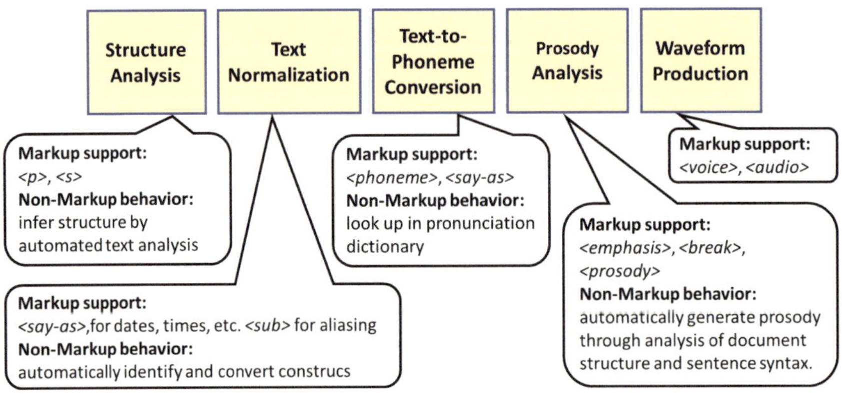

Figure 4.
SSML support for stages of speech synthesis.

```
<?xml version="1.0" encoding="UTF-8"?>
<speak version="1.0"
  xmlns="http://www.w3.org/2001/10/synthesis"
  xml:lang="en-GB">
  <p>The requested flight leaving from
        <s xml:lang="it-IT">
            <sub alias="Roma Fiumicino">FCO</sub></s>
        airport
        <emphasis>with destination
              <sub alias="London Heathrow">LHR</sub>
          </emphasis>
      are: <break time="1s"/>
      <s>
        <sub alias="British Airways 0 3 0 2">BA0302</sub>
        <break time="0.5s"/>
        leaving at
             <say-as interpret-as="time">3:45pm</say-as>
        <break time="0.5s"/>
        from gate number A63.
      </s>
      <!-- Other flight options -->
  </p>
</speak>
```

Figure 5.
A simple SSML document.

3.4 Pronunciation lexicon: PLS 1.0

Both speech grammars and synthesized prompts can require customizing the pronunciation of words in a specific application domain. This is often done by adding a user lexicon. The Pronunciation Lexicon Specification (PLS 1.0 [25]) was created to support the definition of a standard lexicon fully interoperable with SRGS 1.0 and SSML 1.0/1.1. PLS 1.0 became a W3C Recommendation in October 2008.

A PLS document is a container of entries, <lexeme> elements, with a textual part described by the <grapheme> element and with textual replacements provided by <alias> elements or phonetic transcriptions by <phoneme> elements. There can be multiple pronunciations to accommodate different ways to speak a word/token, or for a different spelling for the same pronunciation.

A simple PLS 1.0 document example for a flight application is shown in **Figure 6**. For "Alitalia" and "Lufthansa," the pronunciations inside the <phoneme> element are given in IPA (International Phonetic Alphabet) [34]—a standard way to express the pronunciations for all spoken human languages. Moreover, the two lexemes have a double pronunciation; the first is the normal English one, while the second is closer to their original language (Italian and German, respectively) as spoken by a native speaker of that language. The prefer attribute indicates which pronunciation has to be selected for TTS rendering. For ASR, all the pronunciations will be taken into account.

3.5 Call control—CCXML 1.0

Another language defined by the W3C VBWG targets programming the call control of a voice browser in an innovative way. An XML markup language was developed to define handlers for telephony events generated by a telephone connection or a VoIP SIP interaction. The Voice Browser Call Control (CCXML 1.0) [26] language was designed to allow a very efficient implementation completely based upon events and handlers to avoid creating any latency that might impact the underlying signaling.

```
<?xml version="1.0" encoding="UTF-8"?>
<lexicon version="1.0"
  xmlns="http://www.w3.org/2005/01/pronunciation-lexicon"
  alphabet="ipa" xml:lang="en-GB">
  <lexeme>
    <grapheme>Alitalia</grapheme>
    <phoneme>ˌæl.ɪ.ˈtæi̯ə</phoneme>
    <phoneme prefer="true">a.li.ˈtaː.lja</phoneme>
  </lexeme>
  <lexeme>
    <grapheme>Lufthansa</grapheme>
    <phoneme>ˈlʊft.hænzə</phoneme>
    <phoneme prefer="true">ˈlʊftˌhan.za</phoneme>
  </lexeme>
  <lexeme>
    <grapheme>AF</grapheme>
    <alias>Air France</alias>
  </lexeme>
    <lexeme>
    <grapheme>BA</grapheme>
    <alias>British Airways</alias>
    </lexeme>
</lexicon>
```

Figure 6.
PLS 1.0 document for flight applications.

A CCXML engine is also able to send and receive events through an HTTP/HTTPS connector, which allows for the generation of outbound calls from a Web application and for monitoring calls and conferences via a Web interface.

CCXML 1.0 addresses both simple tasks of call handling (see **Figure** 7), as well as complex ones, such as conditional call handling, conferencing, coaching, etc. Each CCXML document describes transitions to handle specific events. In **Figure** 7, a "connection.alerting" event (incoming call) is accepted by the underlying telephony or VoIP layer, a VoiceXML dialog is started when the "connection.connected" event is received, and then the CCXML processor waits until either the caller disconnects ("connection.disconnect") or the VoiceXML dialog exits ("dialog.exit"). These are simple actions performed during telephony calls, both TDM and VoIP.

While working on the definition of CCXML 1.0, which became a W3C Recommendation in July 2011, the W3C VBWG decided to start another effort to define a state-chart language to generalize the ideas behind CCXML 1.0. This new specification is State Chart XML (SCXML): State Machine Notation for Control Abstraction (SCXML 1.0 [26]), and it can be used as the key component to control a generalized interaction in a multimodal interface. SCXML 1.0 is an XML markup language that provides a generic state-machine-based execution environment inspired by Harel state charts [35].

3.6 IETF protocols: MRCPv1 and v2

The implementation of voice browsing relies on other standards and protocols, the web architecture, with XML documents, namespaces, caching policies to start with, and obviously the HTTP/HTTPS protocols. All these are at the core of the W3C VBWG standards. However, the Internet Task Force Initiative (IETF) [36] was working on needed protocols.

The Media Resource Control Protocol (MRCP), whose initial draft was proposed by CISCO, SpeechWorks, and Nuance, defines the requests, responses, and events to

```
<?xml version="1.0" encoding="UTF-8"?>
<ccxml version="1.0"
   xmlns="http://www.w3.org/2002/09/ccxml">
   <var name="currentState"/>
   <var name="myDialogId"/>
   <var name="myConnId"/>
   <eventprocessor statevariable="currentState">
     <transition event="connection.alerting">
       <assign name="myConnId" expr="event$.connectionid"/>
       <accept connectionid="event$.connectionid"/>
     </transition>
     <transition event="connection.connected">
       <dialogstart
         src="'http://www.example.com/flight.vxml'"
         connectionid="myConnId" dialogid="myDialogId"/>
     </transition>
         <transition event="dialog.started">
         <log expr="'VoiceXML appl is running now'"/>
         </transition>
         <transition event="connection.disconnected">
         <dialogterminate dialogid="myDialogId"/>
         </transition>
         <transition event="dialog.exit">
       <disconnect connectionid="myConnId"/>
         </transition>
         <transition event="*">
         <log expr="'Closing, unexpected:' + event$.name"/>
       <exit/>
     </transition>
   <eventprocessor/>
</ccxml>
```

Figure 7.
Basic handling of incoming calls with CCXML.

control resources of general speech engines, such as ASR and TTS and even speaker verification to enable a distributed and scalable architecture. The initial draft was standardized by IETF as MRCPv1 (RFC 4463 [37]), and it was largely implemented by the industry. The protocol was based on Real-Time Transport Protocol (RTP) for media transport and RTSP (Real Time Streaming Protocol) for controlling speech resources.

In the meantime, standardization continued to MRCPv2, which was instead based on SIP (Session Initiation Protocol) for signaling and SDP (Session Description Protocol) for negotiating and exchanging capabilities. In November 2012, the standardization was completed (RFC 6787 [38]), and it enabled the control of new resources for recording, speaker verification, and identification.

For a complete description of MRCPv2 and its relationship with W3C VBWG standards, see [39].

4. W3C MMIWG standards

The companion working group, Multi-Modal Interaction Working Group (MMIWG), led by Deborah Dahl was attended by almost the same companies attending VBWG. The goal of MMIWG was to extend the scope of standardization beyond the voice or typed input to embrace a much larger set of modalities, such as touch, gesture, emotions, and haptics both as input and output devices for a system.

The major achievements of the W3C MMIWG were the following standards:

- Ink Markup Language, InkML [40], is designed to represent the input of handwriting by a stylus or a finger. In addition to representing traces, InkML offers a rich set of metadata that preserve the appearance of the original input (i.e., color, width, orientation, timing, etc.).

- Extensible multimodal annotation, EMMA [41], is a standard to represent natural language input. It was designed to support annotation from different stages of processing, beginning with the initial results of speech or handwriting recognition and then natural language understanding annotations. EMMA also allows the fusion of different representations across multiple modalities in a multi-modal application, see also [42].

- EMMA was initially inspired by NLSML [43], which is now part of the MRCP protocol, and EMMA 1.0 was then accepted as interpretation result in the MRCPv2 protocol [37].

- Emotion Markup Language, EmotionML [44], is the result of a joint effort of leading researches in the field of emotion and industry. The effort was to create a standard language to annotate emotion in speech, visual, or text corpora, which are not only vital for research but also to represent emotions in recognition engines and to control emotions in TTS rendering. EmotionML became a W3C Recommendation in May 2014. An extended description of EmotionML is available in [45].

Another achievement of the W3C MMIWG was the definition of a multimodal architecture [46]. The multimodal architecture provides an event-based protocol for an interaction manager, possibly implemented in SCXML 1.0 [27], to coordinate an ensemble of modality components, each responsible for processing inputs or producing outputs in specific modalities. The protocol consists of a limited set of standard LifeCycle events—NewContext, Prepare, Start, Pause, Resume, Cancel, Done, ClearContext, Status, and Extension. The standard events include a set of standard fields, for example, fields to record the source and destination of the event, as well as a Data field, which can contain the results of processing an input.

A very detailed and up-to-date description of W3C multimodal standards is contained in [47].

As you see, there was close relationship between these two W3C working groups whose aim was to create a set of interoperable and complementary standards to expand capabilities of state-of-the-art applications.

5. Status and evolutions

This exciting period of an evolution based on standards came to an end after more than 15 years of activity. First, W3C VBWG was declared closed in October 2015 [48] because its mission to support "browsing the Web by voice" was achieved.

Going to the W3C VBWG homepage, there is the list of all the standards created and additional materials (see [13]). The only unfinished work is VoiceXML 3.0 [49] that was the attempt to create an extensible version of VoiceXML where addition of new features would have the benefit of clear interfaces.

When VoiceXML 3.0 effort started, the landscape had changed, greatly due to the success of VoiceXML 2.0 and companion standards. After the adoption of those standards, the industry was in a consolidation process of acquisition of innovative players by larger ones, where the goal was to have those standards at the core of the industry. Therefore, the pressure on innovations was reduced and, as consequence, the process slowed down, and ultimately stopped. One of the last activities was the publication of the first Working Draft of VoiceXML 3.0 [49] before dissolving the working group.

Nevertheless, after more than 20 years, these standards are still firmly at the core of the whole voice application industry, especially for customer care applications and other sectors. The creation of a family of interoperable standards is an advantage, because even new approaches to the development of more advanced speech application, for instance, by hosted APIs [50] or tools, are free to re-use what is already done, such as grammars, TTS controls, lexicons, result formats, and annotations.

A few years later, in February 2017, the W3C MMIWG was also dissolved for similar reasons. The first group of standards that includes InkML 1.0, EMMA 1.0, and EmotionML 1.0 and also the multimodal architecture were completed as W3C Recommendations. Other Working Drafts were also published (see [14]), among them was EMMA 2.0 [51], which was intended to extend EMMA from input results of different modalities to the annotation of output too.

The main lesson learnt is that when times are mature, a neutral and highly collaborative environment, such as W3C working groups, can attract all players that want to innovate to work together for the benefit of a whole industry, or advance new technologies. The shift from proprietary to standard-based technologies was the case described in this paper.

Current human interface platforms are very siloed, using proprietary formats and with little or no concern for interoperability. This means that the kind of inflexible vendor lock-in that we saw 25 years ago with telephony applications is very much with us today. As the underlying technologies continue to evolve, stabilize, and mature, it will become more and more apparent, as it did in the late 1990s, that open standards are a path toward accelerating the ubiquity of voice and multimodal applications and will truly benefit the entire industry.

Acknowledgements

It is thanks to an incredible group of talented people that I wrote this paper. I got to know each of their voices during innumerable conference calls, and their sense of humor in many face2face meetings. First, I would like to thank the chairs, Jim Larson, Dan Burnett, and Debbie Dahl, and not forgetting Scott McGlashan, whom I first met in the early 1990s when he was a PhD student involved in the SUNDIAL project and then later as co-chair with Jim Larson of VBWG. He showed great talent in leading the project's development. His departure in February 2014 was a big loss. I am also indebted to all the people who played such an active role throughout the years. These standards would not have been possible without their efforts. As there are too many to thank individually, I thank them collectively.

The W3C became our home, and from there, I would first like to thank the team contacts who always helped us to understand the W3C's processes and also gave us some very good ideas. Thanks to Dave Raggett, Kazuyuki Ashimura, Max Froumentin, and Matt Womer. Second, thanks go to the other team leads who contributed to broadening the scope of our work, such as Philipp Hoschka, the W3C Domain leader for the Ubiquitous Web, Richard Ishida for Internationalization (I18N), Judy Brewer and Janine Sajka for the Web Accessibility Initiative, and many other great people we met during the W3C Technical Plenary meetings, of course including Tim Berners-Lee.

I also have to mention my second home, the VoiceXML Forum, especially Val Matula and Rob Marchand who sit with me on the Board and, especially, Katie Valenti, our invaluable Program Manager for the ISTO team. Thanks.

I am also very grateful to Debbie Dahl and Roberto Pieraccini who read this paper and contributed so many comments and ideas. Finally, I cannot overlook Simon Parr for his timely assistance.

Author details

Paolo Baggia
Nuance, Torino, Italy

*Address all correspondence to: paolo.baggia@nuance.com

References

[1] Pallett DS. Session 2: DARPA resource management and ATIS benchmark test poster session. In: Proceedings of the 4th DARPA Workshop on Speech and Natural Language. Pacific Grove (CA); 1991. pp. 49-58

[2] Pallett DS, Dahlgren NL, Fiscus JG, Fisher WM, Garofolo JS, Tjaden BC. DARPA February 1992 ATIS benchmark test results. In: Proceedings of the DARPA Speech and Natural Language Workshop. Harriman (NY); 1992. pp. 201-206

[3] Dahl D, Bates M, Brown M, Fisher W, Hunicke-Smith K, Pallett D, et al. Expanding the scope of the {ATIS} task: The {ATIS}-3 corpus. In: Proceedings of Human Language Technology. Plainsboro (NJ); 1994

[4] Paul DB, Baker JM. The design for the Wall Street journal-based CSR corpus. In: Proceedings of the DARPA Speech and Natural Language Workshop. Harriman (NY); 1992. pp. 357-363

[5] Klatt D. Review of text-to-speech conversion for English. The Journal of the Acoustical Society of America. 1987;**82**(3):737-793

[6] Balestri M, Pacchiotti A, Quazza S, Salza PL, Sandri S. Choose the best to modify the least: A new generation concatenative synthesis system. In: Proceedings of the 6th Conference on Speech Communications and Technology (EUROSPEECH-99); 5-9 September 1999; Budapest, Hungary. Vol. 5. 1999. pp. 2291-2294

[7] Peckham J. A new generation of spoken dialog systems: Results and lessons from SUNDIAL project. In: Proceedings of the 3rd Conference on Speech Communications and Technology (EUROSPEECH-93); 22-25 September 1993; Berlin, Germany. 1993. pp. 33-40

[8] Billi R, Castagneri G, Morena D. Field trial of two different information enquiry systems. Speech Communication. 1997;**23**(1-2):83-93

[9] den Os E, Boves L, Lamel L, Baggia P. Overview of the ARISE project. In: Proceedings of the 6th Conference on Speech Communications and Technology (EUROSPEECH-99); 5-9 September 1999; Budapest, Hungary. 1999. pp. 1527-1530

[10] Cole R, Mariani J, Uszkoreit H, Varile GB, Zaenen A, Zampolli A, et al. Survey of the State of the Art in Human Language Technology. Cambridge, UK: Cambridge University Press; 1997

[11] Pieraccini R. The Voice in the Machine, Building Computers that Understand Speech. Cambridge, MA: The MIT Press; 2012

[12] W3C. Voice Browsers W3C Workshop; 13 June 1998; Cambridge, MA. Available from: https://www.w3.org/Voice/1998/Workshop/ [Accessed: 26 April 2020]

[13] W3C. Voice Browser Working Group. Available from: https://www.w3.org/Voice/ [Accessed: 26 April 2020]

[14] W3C. Multimodal Interaction Working Group. Available from: https://www.w3.org/2002/mmi/ [Accessed: 26 April 2020]

[15] Larson J. Introduction and overview of W3C speech interface framework. W3C Working Draft; 4 December 2000. Available from: https://www.w3.org/TR/voice-intro/ [Accessed: 26 April 2020]

[16] Larson JA. W3C speech interface language: VoiceXML. IEEE Signal Processing Magazine. 2007;**4**(3):126-130

[17] VoiceXML Forum. Available from: http://www.voicexml.org/ [Accessed: 26 April 2020]

[18] VoiceXML Forum. Voice eXtensible Markup Language (VoiceXML)

version 1.0. W3C Note; 05 May 2000. Available from: http://www.w3.org/TR/2000/NOTE-voicexml-20000505 [Accessed: 26 April 2020]

[19] McGlashan S, Burnett DC, Carter J, Danielsen P, Ferrans J, Hunt A, et al. Voice Extensible Markup Language (VoiceXML) version 2.0. W3C Recommendation; 16 March 2004. Available from: https://www.w3.org/TR/voicexml20/ [Accessed: 26 April 2020]

[20] Hunt A, McGlashan S. Speech recognition grammar specification version 1.0. W3C Recommendation; 16 March 2004. Available from: https://www.w3.org/TR/speech-grammar/ [Accessed: 26 April 2020]

[21] Burnett DC, Walker MR, Hunt A. Speech Synthesis Markup Language (SSML) version 1.0. W3C Recommendation; 16 March 2004. Available from: https://www.w3.org/TR/speech-synthesis/ [Accessed: 26 April 2020]

[22] Oshry M, Auburn RJ, Baggia P, Bodell M, Burke D, Burnett DC, et al. Voice Extensible Markup Language (VoiceXML) 2.1. W3C Recommendation; 19 June 2007. Available from: https://www.w3.org/TR/voicexml21/ [Accessed: 26 April 2020]

[23] Van Tichelen L, Burke D. Semantic interpretation for speech recognition (SISR) version 1.0. W3C Recommendation; 05 April 2007. Available from: https://www.w3.org/TR/semantic-interpretation/ [Accessed: 26 April 2020]

[24] Burnett DC, Shuang ZW. Speech Synthesis Markup Language (SSML) version 1.1. W3C Recommendation; 07 September 2010. Available from: https://www.w3.org/TR/speech-synthesis11/ [Accessed: 26 April 2020]

[25] Baggia P. Pronunciation Lexicon Specification (PLS) version 1.0. W3C Recommendation; 14 October 2008. Available from: https://www.w3.org/TR/pronunciation-lexicon/ [Accessed: 26 April 2020]

[26] Auburn RJ. Voice browser call control: CCXML version 1.0. W3C Recommendation; 05 July 2011. Available from: https://www.w3.org/TR/ccxml/ [Accessed: 26 April 2020]

[27] Barnett J, Akolkar R, Auburn RJ, Bodell M, Carter J, McGlashan S, et al. State Chart XML (SCXML): State machine notation for control abstraction. W3C Recommendation; 01 September 2015. Available from: https://www.w3.org/TR/scxml/ [Accessed: 26 April 2020]

[28] Barnett J. Introduction to SCXML. In: Dahl DA, editor. Multimodal Interaction with W3C Standards. Cham, Switzerland: Springer International Publishing; 2017. pp. 81-107

[29] Larson J. VoiceXML: Introduction to Developing Speech Applications. Upper Saddle River, New Jersey: Prentice Hall; 2003

[30] Dahl D. Practical Spoken Dialog Systems. Berlin, Heidelberg: Springer-Verlag; 2005

[31] Jokinen K, McTear M. Spoken Dialogue Systems. Princeton, NJ: Morgan & Claypool; 2009

[32] Brown MK, Kellner A, Raggett D. Stochastic language models (N-Gram) specification. W3C Working Draft; 02 January 2001. Available from: http://www.w3.org/TR/ngram-spec [Accessed: 26 April 2020]

[33] Standard ECMA-327. ECMAScript 3rd Edition Compact Profile; June 2001. Available from: http://www.ecma-international.org/publications/files/ECMA-ST-WITHDRAWN/Ecma-327.pdf [Accessed: 26 April 2020]

[34] IPA. Handbook of the International Phonetic Association. Cambridge, UK: Cambridge University Press; 1999

[35] Harel D. Statecharts: A visual formalism for complex systems. Journal Science of Computer Programming. 1987;**8**(3):231-274

[36] The Internet Engineering Task Force (IETF). Available from: https://www.ietf.org/ [Accessed: 26 April 2020]

[37] Shanmugham S, Monaco P, Eberman B. A Media Resource Control Protocol (MRCP). RFC 4463; Informational; April 2006. Available from: https://tools.ietf.org/html/rfc4463 [Accessed: 26 April 2020]

[38] Burnett D, Shanmugham S. Media Resource Control Protocol Version 2 (MRCPv2); RFC 6787; Internet Standard; November 2012. Available from: https://tools.ietf.org/html/rfc6787 [Accessed: 26 April 2020]

[39] Burke D. Speech Processing for IP Networks: Media Resource Control Protocol (MRCP). Chichester, UK: Wiley; 2007

[40] Watt SM, Underhill T. Ink Markup Language (InkML). W3C Recommendation; 20 September 2011. Available from: http://www.w3.org/TR/InkML [Accessed: 26 April 2020]

[41] Johnston M. EMMA: Extensible MultiModal Annotation markup language. W3C Recommendation; 10 February 2009. Available from: http://www.w3.org/TR/emma/ [Accessed: 26 April 2020]

[42] Johnston M. Extensible multimodal annotation for intelligent interactive systems. In: Dahl DA, editor. Multimodal Interaction with W3C Standards. Cham, Switzerland: Springer International Publishing; 2017. pp. 37-64

[43] Dahl DA. Natural language semantics markup language for the speech interface framework. W3C Working Draft; 20 November 2000. Available from: https://www.w3.org/TR/nl-spec/ [Accessed: 26 April 2020]

[44] Burkhardt F, Schröder M. Emotion Markup Language (EmotionML) 1.0. W3C Recommendation; 22 May 2014. Available from: https://www.w3.org/TR/emotionml/ [Accessed: 26 April 2020]

[45] Burkhardt F, Pelachaud C, Schuller BW, Zovato E. EmotionML. In: Dahl DA, editor. Multimodal Interaction with W3C Standards. Cham, Switzerland: Springer International Publishing; 2017. pp. 65-80

[46] Barnett J, Bodell M, Dahl D, Kliche I, Larson J, Porter B, et al. Multimodal architecture and interfaces. W3C Recommendation; 25 October 2012. Available from: https://www.w3.org/TR/mmi-arch/ [Accessed: 26 April 2020]

[47] Dahl DA, editor. Multimodal Interaction with W3C Standards. Cham, Switzerland: Springer International Publishing; 2017

[48] Burnett DC. ALL: Thoughts and thanks as the VBWG comes to a close. W3C Mailing List Archive; 26 September 2015. Available from: https://lists.w3.org/Archives/Public/www-voice/2015JulSep/0029.html [Accessed: 26 April 2020]

[49] McGlashan S, Burnett D, Akolkar R, Auburn RJ, Baggia P, Barnett J, et al. Voice Extensible Markup Language (VoiceXML) 3.0. W3C Working Draft; 16 December 2010. Available from: http://www.w3.org/TR/voicexml30/ [Accessed: 26 April 2020]

[50] Natal A, Shires G, Cáceres M, Jägenstedt P. Web speech API. Draft Community Group Report; 21 January 2021. Available from: https://wicg.github.io/speech-api/ [Accessed: 26 April 2020]

[51] Johnston M, Dahl DA, Denney T, Kharidi N. EMMA: Extensible MultiModal Annotation markup language version 2.0. W3C Working Draft; 08 September 2015. Available from: https://www.w3.org/TR/2015/WD-emma20-20150908/ [Accessed: 26 April 2020]

Section 3

Cognitive Processing

Chapter 5

The Influence of Cognitive Biases in Production Planning and Control: Considering the Human Factor for the Design of Decision Support Systems

Julia Bendul and Melanie Zahner

Abstract

Production planning and control (PPC) requires human decision-making in several process steps like production program planning, production data management, and performance measurement. Thereby, human decisions are often biased leading to an aggravation of logistic performance. Exemplary, the lead time syndrome (LTS) shows this connection. While production planners aim to improve due date reliability by updating planned lead times, the result is actually a decreasing due date reliability. In current research in the field of production logistics, the impact of cognitive biases on the decision-making process in production planning and control remains at a silent place. We aim to close this research gap by combining a systematic literature review on behavioral operation management and cognitive biases with a case study from the steel industry to show the influence of cognitive biases on human decision-making in production planning and the impact on logistic performance. The result is the definition of guidelines considering human behavior for the design of decision support systems to improve logistic performance.

Keywords: cognitive biases, human behavior, production planning and control, PPC, system design

1. Introduction

While in the area of psychology, anthropology, and sociology human behavior has been investigated intensely and although behavioral aspects became an integral part of economic research [1], in the field of logistics and production planning and control (PPC), only little research has been conducted [2]. In order to support decision-making processes in PPC and to optimize logistic performance, various models have been developed in order to reach short lead times, high due date reliability, low inventory levels, and high-capacity utilization as the key logistic performance indicators for production systems [3]. However, the underlying proposition of these models is typically the theory of the "homo

IntechOpen

economicus." In other words, to apply these models properly, we assume a fully rational human behavior in the decision-making process determined by the purpose of the decision-maker to maximize the personal advantage [4]. Tversky and Kahneman [5] challenged this assumption and showed that human decisions are biased, which means a systematic deviation from rational judgment. In the fields of logistics and PPC, people are often confronted with uncertainty and high complexity, and research has shown that under these framework conditions, humans systematically take wrong decisions [6]. One example for a complex situation in which biased decision-making leads to a deteriorating logistic performance is the so-called lead time syndrome (LTS). Here, production planners overreact to decreasing due date reliability. The planners adapt standard lead times too often, which eventually leads to an even worse aggravation of due date reliability [7]. To support this variety of decisions, which have to be made in PPC, the so-called decision support systems (DSSs) are used frequently. DSSs are computer-based information systems with the purpose to improve the decision-making process and its outcome [8].

In this chapter we aim to improve the understanding of the role of cognitive biases in the field of PPC and propose first design guidelines for decision support systems (DSSs). Therefore, we combine a systematic literature review on behavioral operation management and cognitive biases. Taking inspiration from a case study from the steel industry, we show the possible impact of cognitive biases on human decision-making in PPC and on logistic performance. The remainder of this chapter is structured as follows. In Section 2, we outline the typical decision-making processes and the corresponding DSS in PPC. In Section 3, we use the case of the PPC at a steel manufacturer to present several examples of the possible impacts of cognitive biases on PPC decision-making. In Section 4, we give first recommendations on how to avoid the emergence of cognitive biases within PPC decision-making and derive first guidelines for DSS.

2. Decision-making and decision support systems in PPC

2.1 Decision-making in operation management

Already in the 1980s, the *decision-making* has been recognized as a field of central research interest in the area of operation management [8]. Decision-making can be defined "[...]as the process of selecting the course of action that best meets the decision criteria, subject of the constraints inherent in the decision-making situation (DMS)." [9] p. 324.

According to [8] the DMP contains three phases. In the first "intelligence" phase, the problem which requires a solution by the decision-maker is identified and prioritized. Moreover, the first target achievements are defined, and corresponding data gathering is initialized. In the second "design" phase, a general action plan, which contains several action alternatives and their expected outcomes as well as the first evaluation criteria, is defined. In the third "choice" phase, the decision-maker selects the best action alternative based on the evaluation of each alternative. Based on the early works of [8], several models have been developed in order to explain the DMP. For instance, several authors suggest an extension of [8] DMP model [9, 10]. They propose a fourth "implementation" phase in which the decision outcome is turned into practice. In a fifth "learning" phase, lessons learned are formulated and shared in the organization to improve the DMP and the decision outcome in the future.

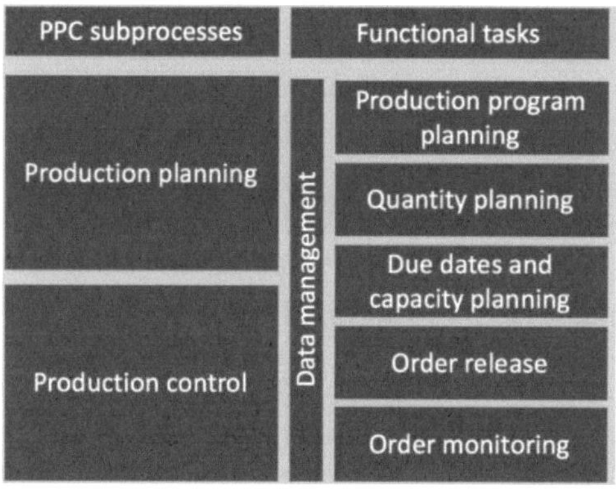

Figure 1.
PPC model according to [11].

2.2 Decision-making in production planning and control

Moreover, also, within PPC human decisions undergo the suggested phases of the DMP models. Typical decision-making situations of PPC are shown in **Figure 1**.

PPC contains two subprocesses which are production planning and production control. These two subprocesses contain several task functions which require several decisions. Production planning focuses on the development of the basic concept to determine when to produce what in which quality. Production control has an overall monitoring function to achieve the production targets by the use of different control techniques.

P*roduction program planning* encompasses several decisions about the production sequence and the required materials. Based on this, *quantity planning* determines production quantities and lot sizes. *Due dates and capacity planning* contain several decisions concerning capacity plans and due dates for specific production steps. The *order release* marks the starting point for the production. Since disturbances, such as machine breakdowns, delays in material supply, or quality problems, may occur during production, a continuous *order monitoring* is accomplished. The necessary decisions within these main tasks of PPC are often complex and require the consideration of several parameters. Thus, in PPC typically decision-making is usually supported by DSS. DSSs are often self-developed by using case tools like Crystal, Analytica, or iThink for the development.

2.3 Insufficient design guidelines for decision support systems

While there is a lot of research on DSSs in general (e.g., [12]), as well as on several components and modules of DSS (e.g., [13]), there is only little research about standardized design guidelines for DSS. For instance, [6, 9] criticize the lack of an integrated framework to support a standardized design of DSS. However, [9] propose a framework as the basis for standardized guidelines for DSS designers.

Figure 2 shows the framework suggested by [9]. The framework contains four levels. (1) The first *decision-making level* is based on the original DMP model of [8] and on its extensions containing all five steps within the decision process. (2) The second *decision service task level* focuses on tasks which require human intelligence and is based on a task-method-subtask structure to infer logical conclusions from the analyzed data. (3) The third *architectural capability level* considers user interface, data information knowledge, and processing capabilities. (4) The fourth *computational symbol-program level*

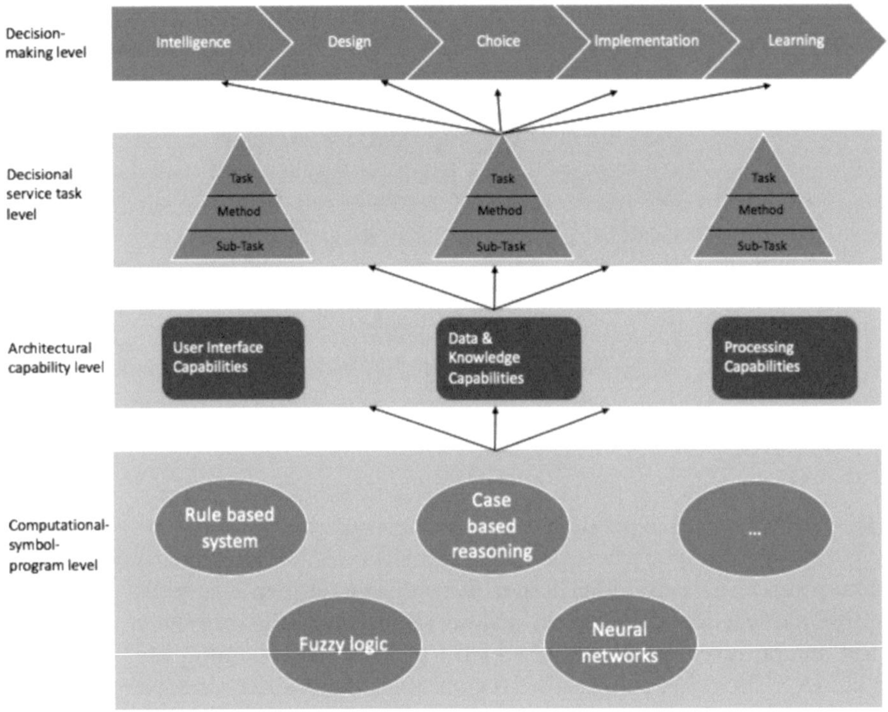

Figure 2.
DSS design framework with adaption according to [9].

focuses on specific computational mechanisms based on artificial intelligence techniques such as computer-based reasoning (CBR), rule-based system (RBS), etc.

3. Cognitive biases in production planning and control: the case of a German steel producer

3.1 Foundations of research on cognitive biases

Tversky and Kahneman [5] were the first who questioned the assumption of rational human behavior and introduced the term of cognitive biases. They state that humans taking decisions systematically go wrong, especially in complex and uncertain environments. In further experiments, [14] deepens this research of the underlying factors and describes the cognitive processes of intuition and reasoning.

Stanovich and West [15] named these cognitive processes System I (intuition) and System II (reasoning). While System I acts automatically, fast, emotively, and effortlessly and is hardly controllable, System II operates relatively slowly, reflected, and effortful [15]. System I creates spontaneous impressions and persuasions, which form the basis for further decisions and actions of System II. Based on this two-system view, [14] claims that impressions are generated in System I and judgments are made in System II.

This fundamental research made clear that there are plenty of different cognitive biases that may affect human decision-making. Ref. [6] categorized these biases into six main categories:

1. Memory biases describe biases influencing the storage and the ability to remember information.

2. Statistical biases are the human tendency to over- or underestimate certain statistical parameter.

3. Confidence biases act to increase a person's confidence in their prowess as a decision-maker.

4. Adjustment biases describe the human tendency to stick to the first available information or to a reference point when making decisions.

5. Presentation biases influence humans in their decision-making by the way how information is being displayed.

6. Situation biases describe the way how a person responds to the general decision situation.

3.2 Case study: decreasing due date reliability at a German steel producer

3.2.1 Initial situation

We take inspiration from a case study of the steel industry presented by [2]. The analyzed PPC process takes place within a R&D department of a German steel case company. To compete in the global steel market, a short time to market is crucial. Especially in the R&D department, production and analysis processes are hardly to plan, and it is one of the major challenges of production planners to fulfill the customer requested delivery date. Samples of new alloys have to pass sequences of different tests before they can be launched in the market. In the analyzed R&D process, the first orders for different steel samples are placed through external and internal customers. After estimating the planned lead time for several manufacturing and analysis processes, the orders get a due date. For the scheduling of the production orders, a custom-developed DSS is used. In total, a production system with 20 machines and 35 employees in 1 shift was analyzed over a period of 3 years (from 2011 to 2014, 1.023 orders were analyzed). On-site visits, expert interviews, and observation documents were the used research methods. To evaluate the key performance indicator (KPI) development in terms of due date reliability, inventory, and lead times, feedback data from 13 months based on 240 shop floor calendar days were analyzed.

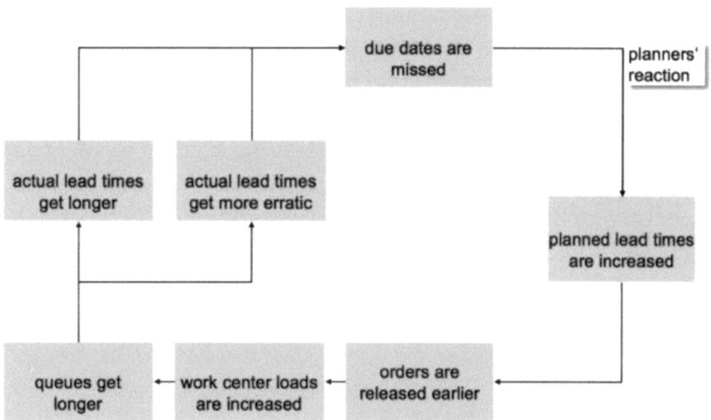

Figure 3.
Lead time syndrome in PPC.

3.2.2 Observed behavior of key performance indicators

The due date reliability was one of the key performance indicators, and 95% was set as a long-term target for the planners. We observed the so-called lead time syndrome active in this context. When planners recognized decreasing delivery reliability, they started to update the initially planned lead times by releasing waiting orders earlier and adding some additional safety lead times in that cases in which the initially lead time was too short to meet the target due date. Thus, more orders are in the production system which causes an increasing WIP level and growing lead times. As a result, the delivery reliability was even lower than before the update of the lead times. The planners feel pressured to improve the current situation, and the circle of updating lead times reinforces, resulting in an even stronger due date aggravation.

Figure 3 shows the process of the observed planner's behavior.

3.2.3 Observed behavior of planners: cognitive biases underlying deteriorating due date reliability

We observed several active cognitive biases in various decision-making situations in several PPC tasks. Nevertheless, it is important to understand that this classification of biases is not as concrete in practice as described in theory. Some of the cognitive biases overlap and often occur in several different situations.

3.2.3.1 Memory biases

Memory biases summarize a group of cognitive biases which are related to the storage and availability of information. The *availability heuristic* describes the tendency of people to overestimate the likelihood of events for which they can easily restore the information [14]. As a result, people tend to overweight the outcome of the last decision as a basis for their decision-making in their current situation. The *imaginability bias* describes the fact that people assume an event to be more probable if it can be easily imagined by themselves [16].

We observe that planners tend to adjust planned lead times based on their intuition instead of entirely considering all influencing variables.

This occurs mainly in the phase of the *production program planning* and *due date and capacity planning*. The planners tend to connect their last updates of the planned lead time to any positive development of the logistic performance. In case of a negative development, the planners assume that external influences such as a delay in material delivery or machine failures are responsible for the fact delayed due date reliability. They conclude that there is a need for another planned lead time adjustment.

We observe the same development for the *production program planning*. In the cases when the defined production sequence leads to a positive performance outcome, the planners relate this development to the quality of their own planning capabilities. In those cases when the defined production sequence leads to a decreasing due date reliability, the planners connected this with external influencing factors and conclude a necessary update on the production sequence, even though this was not optimal for the current situation.

3.2.3.2 Statistical biases

Statistical biases describe the tendency to over- or underestimate certain statistical parameters. Ref. [14] investigates that humans tend to overestimate the probability of two events occurring together if this has already happened once in the past. This effect is described by the *correlation bias*. For example, a change in material and an increase in lead times for a certain machine can lead to the assumption that there is a correlation between this material change and extending lead times—which actually does not exist. The *gambler's fallacy* describes the phenomenon of the assumption that future events are determined by the occurrence of past events [17]. This leads to an overestimation of possible events ignoring the actual statistical possibility [18].

We observe statistical biases in the adaption of lead times within the *phase of the order monitoring*. The planners tend to assume that the coincident adaption of planned lead times and the positive development of due date reliability are correlated, although they are aware of the mathematical fact that it takes 4 weeks until the adaptation of planned lead times will become visible in an improved due date reliability.

We observe these biases also in the *phase of production quantity planning and the order release*. Delays in material provision and machine breakdowns which occurred at the same time lead to the assumption that there is a possible correlation and that this may also increase the system's scrap rate. However, the planners do not further validate this assumption, and the planners simply increase the material orders to reach the desired production output. As a result, the inventory level increases excessively, since the additionally ordered material cannot flow off because the assumed correlation was not true.

3.2.3.3 Confidence biases

Confidence biases describe the set of biases concerning the person's confidence in their prowess as a decision-maker. The *illusion of control or overconfidence biases* describes the tendency of people to overestimate their ability to solve difficult problems [19]. The *conformation bias* leads people to seek for information which confirms their own estimation and hides information which is contrary to their own perception [20].

Analyzing the case study, we find three examples of the *illusion of control* in the *phase of due date planning*. (1) Planners tend to assume that their own procedures are more suitable than the standard planning procedures. (2) When planning the lead times, they behave as if the stable forecast of future incoming orders is predetermined and not only a prediction. (3) Planners increased the WIP level via the planned lead times in order to avoid the production system to run into an idle state. We also found situations exemplary for active confirmation biases. Planners let themselves be guided by their intuition: if planners *feel* that updating the lead times would be the best option to increase due date reliability, they search particularly for information which confirms this feeling. Obvious information which entails the result not to intervene in the planned lead times (such as the given *planning rules* that limit the number of planned lead time updates within a certain period of time) is ignored.

Confirmation biases were also observed in the phases of *production program planning* and *production quantity planning*. Planners behave as if the estimated future customer demand forecasts are stable and the demand numbers are already fixed. Accordingly, they ordered the corresponding materials and plan production sequences without any buffers accordingly. Moreover, we notice that even when the customer demand is in the course of time and can be determined more specifically (no matter whether it is higher or lower than previous forecasts), planners seek for information which confirms the first numbers in order to justify that they stick to their initial plan (e.g., they search for cases in which certain customers have increased order quantities at first and then lowered them).

3.2.3.4 Adjustment biases

Adjustment biases describe the human tendency to stick to the first available information or to a reference point when making decision. The *anchoring effect* is defined as the tendency to rely on an initially given information too heavily—which influences further values [5]. Teng and Das [21] show that anchoring can create systematic errors in decision-making situations. Adjustment biases also include *conservatism bias*. Similar to the anchoring effect, taken estimations are not updated according to new information [22].

This became obvious in the phase of the *order release* and *order monitoring* in the context of lead times. We find that lead times from previous years and from similar work systems act as anchors. When setting planned lead times, planners justify the extension of planned lead times with the numbers in the year 2014. Similarly, the planners tend to aim at a due date reliability of 95%, which is given as the long-term goal (but which is far from reality), and therefore seem to extend the planned lead times disproportionately.

The same effect becomes obvious for the capacity planning. The planners justify their machine capacity planning with target figures of machine utilization rates of the previous years. These figures were not updated to the current situation.

3.2.3.5 Presentation biases

Presentation biases summarize a set of cognitive biases which influence humans in their decision-making by the way how information is being displayed. The *ambiguity effect* describes the human tendency to favor simple-looking options and avoid options that seem to be complicated [23]. According to the *primacy/recency effect,* information at the beginning and at the end of a series can be restored best, whereas information in the middle are restored worst [24].

The implemented DSS offers plenty of types of analysis (such as the order forecasts, inventory levels, etc.) next to the information which is central for setting planned lead times. The *primacy recency effect* became obvious in the phase of order monitoring when the planner was setting the planned lead time for a certain order to the double value of what was reasonable. This is because he had just checked the current due date reliability and noticed that the value for the previous day was particularly low.

Further, we identified the influence of the ambiguity effect in the material quantity planning. When planners recognized that the production could run out of material, they just increased the initially ordered quantity. They did not further analyze potential causes like an increasing scarp rate due to an incorrectly set machine, etc. Instead they took the simplest option right in front of them to keep the production running even though the failure cause exponentiated.

3.2.3.6 Situation biases

Situation biases describe the way how a person responds to the general decision situation. The *complexity effect* describes that people are biased under time pressure or when information overload occurs [25]. The *ostrich effect* describes the habit of people to ignore an obvious negative information [26]. The *bandwagon effect* describes the tendency to do things because many other people do the same [4].

We identify situation biases in several tasks in PPC. Modern PPC DSSs provide a wide range of information, such as key performance indicators concerning delivery reliability, inventory levels, or throughput times. For many planners, the amount and variety of information are too much to be included in their

	Memory Biases	Statistical Biases	Confidence Biases	Adjustment Biases	Presentation Biases	Situation Biases
Production program planning	▪ Availability heuristic ▪ Imaginability bias		▪ Illusion of control ▪ confirmation bias			▪ Complexity effect ▪ Bandwagon effect ▪ Ostrich effect
Quantity planning		▪ Gambler's fallacy ▪ Correlation bias	▪ Illusion of control ▪ confirmation bias		▪ Ambiguity effect ▪ Primacy recency effect	
Due dates and capacity planning	▪ Availability heuristic ▪ Imaginability bias		▪ Illusion of control ▪ confirmation bias	▪ Anchoring effect ▪ Conservatism bias		▪ Complexity effect ▪ Bandwagon effect ▪ Ostrich effect
Order release		▪ Gambler's fallacy ▪ Correlation bias		▪ Anchoring effect ▪ Conservatism bias		
Order monitoring		▪ Gambler's fallacy ▪ Correlation bias	▪ Illusion of control ▪ confirmation bias		▪ Ambiguity effect ▪ Primacy recency effect	▪ Complexity effect ▪ Bandwagon effect ▪ Ostrich effect

Figure 4.
Observed cognitive biases in the case of steel production PPC.

decision-making. In particular, under time pressure the planners decide to extend lead times just to do anything, even when they do not come to a reasonable conclusion when analyzing the data. At the same time, the planners ignore the fact that their own behavior of extending lead times influences due date reliability in a negative way. Moreover, we find that adjusting lead times is a common method of reacting to decreasing due date reliability. Planners who face the situation of decreasing due date reliability choose planned lead time extension just because their colleagues do so. Also, in the case of a machine breakdown, a similar behavior could be observed. The closer the delivery due date, the more planners decided to switch machines and change the production program sequence just to do anything. This was even true when the effort and time to change machines take in total longer than the repair of the initial machine.

Figure 4 shows a summary of our observed cognitive bias categories within the several PPC tasks. Potentially, there are even more active biases in the several PPC tasks, which we did not observe in our case.

4. Debiasing by the design of decision support systems

DSSs intend to improve the decision outcome by supporting the human decision-making process [6]. Therefore, in the design of DSSs, also human behavioral aspects need to be considered to get an unbiased decision outcome. Based on our identified cognitive biases in PPC decisions, we aim to give first recommendations for system developers of DSS.

The proposed framework of [9] serves as the basis and is extended by a so-called behavioral layer. In this, already in the design phase, the DSS should foresee adequate *debiasing techniques* to support planners properly and thus to positively affect logistic performance of the production system.

Debiasing is a method to reduce or eliminate the influence of cognitive biases within the decision process. Keren [27] proposed the following three steps for effective debiasing:

1. Identification of the existence and nature of the potential bias and the underlying influencing factors

2. Consideration of ways and techniques to lower the impact of bias

3. Monitoring and evaluation of the effectiveness of the selected debiasing technique

The proposed steps should be included in the design of a DSS. Based on our findings about the active cognitive biases in PPC, we already fulfilled the first step. In this section we contribute to the second step and aim to propose ways and techniques to lower the impact of biases. The third step then needs to be analyzed and observed over time.

These steps form the generic basis for a debiasing approach which contains further debiasing categories describing the concrete method of debiasing.

Kaufmann et al. [28] propose five categories for effective debiasing strategies in supplier selection processes:

1. Decomposing/restructuring

2. Put yourself in the shoes of

3. Draw attention to alternative outcome

4. Devil's advocate

5. General bias awareness

We used these categories as a basis for the development of the first design guidelines for DSS in the field of PPC.

(1) Decomposing/restructuring: By applying this method, the decision and the related information are restructured and separated to match the task and the capabilities of the decision-maker [29]. Therefore, decisions, such as production program planning within PPC, should be split into single tasks. In other words, the production program planning should be broken down into the subtasks of the decision about the production of several product categories, the corresponding quantities, and the due dates. In order to avoid an information overload for the DSS user and the occurrence of the situation biases (which may cause losing the overview of the connection between single parameters), the DSS should only show the most relevant information for a decision. Additional parameters should be available in the system in the background and should be provided upon request. For example, a machine breakdown can entail a decision update on the capacity planning because the production quantity originally planned on the broken machine has to be switched to another machine. Occasionally, this can also result in a necessary update of the production program. If this is the case, only in the moment when a decision becomes relevant, the request should be provided.

(2) Put yourself in the shoes of: The objective of this method is to enable the decision-maker to consider all the influencing parameters of affected parties through a perspective shift [28]. To enable this method, a pre-analysis of the affected stakeholders of potential decision cases is required.

Within the DSS this method can be applied in two ways. (1) First, before making a final decision, potential scenarios can be presented to the decision-maker. This should also include the implications for logistic performance (e.g., inventory levels, lead time, due date reliability). For example, the implications of a change the production program may have for machine and personnel capacity as well as for material requirement planning should be made visible to the planner even if these implications are only of interest for other departments (e.g., the logistics and the purchasing department). (2) Second, historical data on the decision-making and the outcomes in terms of logistic performance can be provided (e.g., the decision about the lead time in the previous month caused this delivery reliability).

(3) Draw attention to alternative outcome: This method focuses on alternative outcomes to avoid the confirmation biases seeking for supportive information on the initial hypotheses. Thinking about counter explanations as well as the opposite intention and perspective can broaden one's own decision-making horizon. For example, in the case of a machine failure, the first intention of planners in our observed case was to switch the machines to stick to the production due date. However, this caused additional setup time. The opposite intention here would be to stick to the initial planned machine and wait for the machine to be repaired or start with some tasks which can be done without the machine to avoid losing time due to the machine breakdown and avoiding additional setup time at the same time.

(4) Devil's advocate: This debiasing strategy focuses on the possible critique of other parties affected by the taken decision. Thereby, the *devil's advocate* argues against the position of the decision-maker. Through this presentation of a formalized dissent, the decision-maker is forced to proof his decision and find supportive arguments. Research has shown that this leads to better solutions [28, 30]. An important criterion to apply this method successfully is that the devil's advocate is nonemotional in raising his dissent [31]. Therefore, including this method in a DSS is appropriate to fulfill this criterion. Before executing the final decision regarding the extension of a planned lead time, a pop-up window should arise and present a summary of all possible negative effects linked to the question whether the decision-maker is sure about continuing with his decision. Exemplarily, in a case of intended machine switch which also causes setup requirements for tooling, etc., the system should ask whether this really should be done.

(5) General bias awareness: The general awareness of the existence of cognitive biases can be understood as an overall debiasing strategy. Even if the general understanding of the underlying influencing factors on decisions can improve decisional judgment quality, it cannot completely eliminate its emergence [32]. A wider understanding of the influence of cognitive biases on decision-making can be achieved, for example, by provision of short training videos or a user tutorial explaining the influence of cognitive biases. This can be the starting point before using the DSS tool initially. The *general bias awareness* can also be affected by the layout of the graphical user interface as well as by the structure of the DSS which should be well organized and intuitively understandable. This contributes to the avoidance of *situation biases* due to an information overload. Moreover, in attention to the *statistical and the anchoring biases*, just presenting simple figures should be avoided, and additional context information should be added. Based on our observed case study of the long-term target delivery reliability of 95% which acted as an anchor and was quite unrealistic, it would be better to give additional information such as a delivery reliability target for each month and more content information about corresponding developments such as an allowed inventory level or machine utilization rate to achieve this goal.

Figure 5 shows our proposition for a further design layer for the DSS design framework presented by [9].

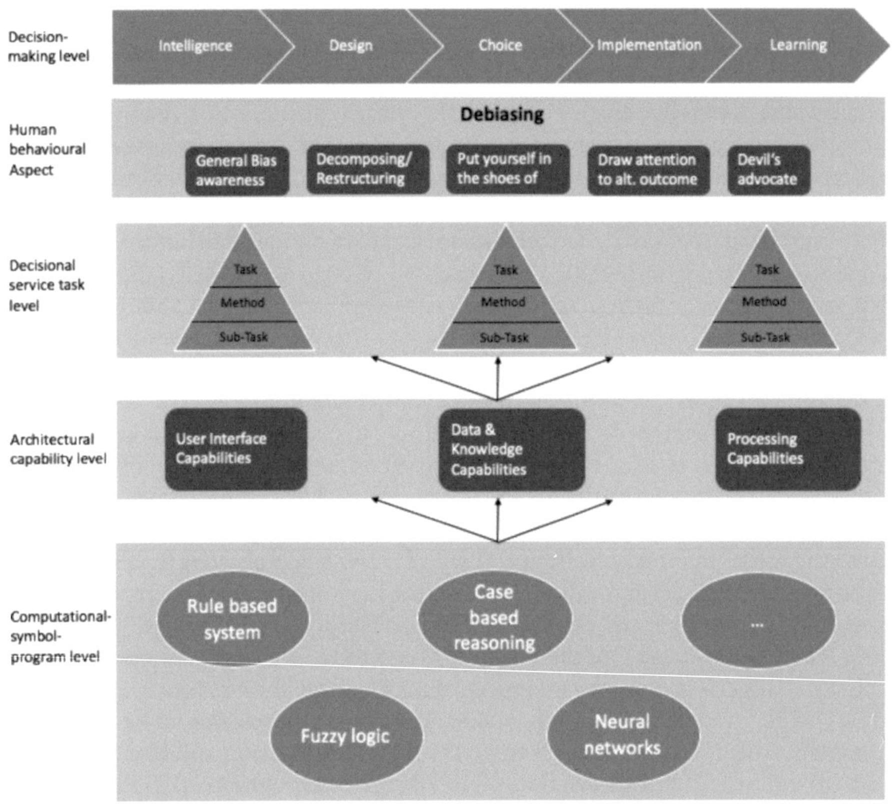

Figure 5.
DSS design framework presented by [9] with our proposition of an additional layer for the DSS design framework.

5. Conclusion

Behavioral aspects in operations management have been investigated for several years. We contribute to this research stream by analyzing the meaning of cognitive biases for decision-making in the field of PPC. The aim of this chapter is to determine first design guidelines for DSS considering the behavioral factors influencing human decision-making. The presented case study shows this need for industrial practice. Frameworks that aim to give advice to designers of DSS ignore the importance of the human factor in decision-making. We contributed to this research by extending the proposed design framework of [9] by adding the human behavioral layer. Moreover, we show first possible design techniques for DSS considering debiasing methods especially for decisions in PPC.

Author details

Julia Bendul* and Melanie Zahner
RWTH Aachen University, Germany

*Address all correspondence to: bendul@scm.rwth-aachen.de

References

[1] Tokar T. Behavioural research in logistics and supply chain management. International Journal of Logistics Management. 2010;**21**:89-103. DOI: 10.1108/09574091011042197

[2] Bendul J, Knollmann M. The human factor in production planning and control: Considering human needs in computer aided decision-support systems. International Journal of Manufacturing Technology and Management. 2015;**30**:346-368. DOI: 10.1504/IJMTM.2016.078921

[3] Nyhuis P, Wiendahl H-P. Fundamentals of Production Logistics: Theory, Tools and Applications. Berlin, Heidelberg, New York: Springer; 2009. DOI: 10.1007/978-3-540-34211-3

[4] Carter C, Kaufmann L, Michel A. Behavioral supply management: A taxonomy of judgment and decision making biases. International Journal of Physical Distribution and Logistics Management. 2007;**37**:631-669. DOI: 10.1108/09600030710825694

[5] Tversky A, Kahneman D. Judgment under uncertainty: Heuristics and biases. Science. 1974;**18**:1124-1131. DOI: 10.1126/science.185.4157.1124

[6] Arnott D. Cognitive biases and decision support systems development: A design science approach. Information Systems Journal. 2006;**16**:55-78. DOI: 10.1111/j.1365-2575.2006.00208

[7] Arnott D. Decision support systems evolution: Framework, case study and research agenda. European Journal of Information Systems. 2004;**13**:247-259. DOI: 10.1057/palgrave.ejis.3000509

[8] Simon HA, Dantzing G, Hogart R, Plott C, Raiffa H, Schelling T, et al. Decision making and problem solving. Interfaces. 1987;**17**:11-31. DOI: 10.1287/inte.17.5.11

[9] Mora M, Cervantes F, Garrido L, Gupta J, Gelman O. Toward a comprehensive framework for the design and evaluation of intelligent decision-making support systems (i-DMSS). Journal of Decision Systems. 2005;**14**:321-344. DOI: 10.3166/jds.14.321-344

[10] Rowe J, Davis A. Intelligent Information Systems. Westport: Quorum; 1996

[11] Hackstein R. Produktionsplanung und -steuerung (PPS) – Ein Handbuch für die Betriebspraxis. Berlin: Springer; 1984. p. 372

[12] Sprague RH. A framework for the development of decision support systems. Management Information Systems Quarterly. 1980;**4**:1-26. DOI: 10.2307/248957

[13] Baldwin J, Allen P, Ridgway K. An evolutionary complex systems decision-support tool for the management of operations. International Journal of Operations & Production Management. 2010;**30**:700-720. DOI: 10.1108/01443571011057308

[14] Kahneman D. Maps of bounded rationality: A perspective on intuitive judgment and choice. Nobel Prize Lecture. 2002:449-489. DOI: 10.1257/000282803322655392

[15] Stanovich KE, West RF. Individual differences in reasoning: Implications for the rationality debate? The Behavioral and Brain Sciences. 2000;**23**:645-726. DOI: 10.1017/S0140525X00003435

[16] Taylor SE, Thompson SC. Stalking the elusive 'vividness' effect. Psychological Review. 1982;**89**:155-181. DOI: 10.1037/0033-295X.89.2.155

[17] Barron G, Leider S. The role of experience in the Gambler's fallacy.

Journal of Behavioral Decision Making. 2010;**23**:117-119. DOI: 10.1002/bdm.676

[18] Hogarth RM. Judgment and Choice: The Psychology of Decision. 2nd ed. Oxford: Wiley; 1987. ISBN: 0471914797

[19] Brenner LA, Koehler DJ, Liberman V, Tversky A. Overconfidence in probability and frequency judgements: A critical examination. Organisational Behaviour and Human Decision Processes. 1996;**65**:212-219. DOI: 10.1006/obhd.1996.0021

[20] Russo JE, Medvec VH, Meloy MG. The distortion of information during decisions. Organizational Behaviour and Human Decision Processes. 1996;**66**:102-110. DOI: 10.1006/obhd.1996.0041

[21] Teng B-S, Das TK. Cognitive biases and strategic decision processes. An integrative perspective. Journal of Management Studies. 1999;**36**:757-778. DOI: 10.1111/1467-6486.00157

[22] Nelson MW. Context and the inverse base rate effect. Journal of Behavioral Decision Making. 1996;**9**:23-40. DOI: 10.1002/(SICI)1099-0771

[23] Ellsberg D. Risk, ambiguity and the savage axioms. Quarterly Journal of Economics. 1961;**75**:643-669. DOI: 10.2307/1884324

[24] Yates JF, Curley SP. Contingency judgement: Primacy effects and attention decrement. Acta Psychologica. 1986;**62**:293-202. DOI: 10.1177/0272989X8800800310

[25] Ordonez L, Benson L. Decisions under time pressure: How time constraint affects risky decision making. Organizational Behaviour and Human Decision Processes. 1997;**71**:121-140. DOI: 10.1006/obhd.1997.2717

[26] Karlsson N, Loewenstein G, Seppi D. The ostrich effect: Selective attention to information. Journal of Risk and Uncertainty. 2009;**38**:95-115. DOI: 10.1007/s11166-009-9060-6

[27] Keren G. Cognitive aids and debiasing methods: Can cognitive pills cure cognitive ills. In: Caverni JP, Fabre JM, Gonzalez M, editors. Cognitive Biases. Amsterdam: North-Holland; 1990. pp. 523-555

[28] Kaufmann L, Carter C, Buhrmann C.Debiasing the supplier selection decision: A taxonomy and conceptualization. International Journal of Physical Distribution and Logistics Management. 2010;**40**:792-821. DOI: 10.1108/09600031011093214

[29] Ashton RH, Kennedy J. Eliminating recency with self-review: The case of auditors' 'going concern' judgments. Journal of Behavioral Decision Making. 2002;**15**:221-231. DOI: 10.1002/bdm.412

[30] Valacich JS, Schwenk CR. Devil's advocacy and dialectical inquiry effects on face-to-face and computer-mediated group decision making. Organizational Behavior and Human Decision Processes. 1995;**63**:158-173. DOI: 10.1006/obhd.1995.1070

[31] Schwenk CR, Cosier RA. Effects of the expert, devil's advocate, and dialectical inquiry methods on prediction performance. Organizational Behavior and Human Performance. 1980;**26**:409-424. DOI: 10.1016/0030-5073(80)90076-8

[32] Babcock L, Loewenstein G. Explaining bargaining impasse: The role of self-serving biases. The Journal of Economic Perspectives. 1997;**11**:109-126. DOI: 10.1257/jep.11.1.109

Chapter 6

Towards Agility and Speed in Enriched UX Evaluation Projects

Juliana Alvarez, David Brieugne, Pierre-Majorique Léger, Sylvain Sénécal and Marc Frédette

Abstract

Recent research has called for the use of enriched measures, that is, psychophysiological measures of emotional and cognitive states, in user experience (UX) testing. This chapter investigates how these enriched measures can inform user experience evaluation while maintaining agility and speed in managing UX projects. Using a multiple case approach, this chapter presents the analysis of 12 recent user experience projects in which enriched measures were used. Lessons learned with regard to challenges encountered are outlined. They emphasize on: (1) the nature of the research question impacts the completion time and the complexity of the project; (2) the need to communicate and coordinate between all parties; (3) the need to anticipate the collected measurements and enhanced results using a mosaic of hybrid collection methods; (4) the nature of the results adapted to underline the operational side without reducing the quality of the work performed; and (5) the time constrains influenced and influencing the pre-tests and project's granularity. This chapter concludes with lessons learned from an agile/UX development approach in the realization of *Sprint projects*.

Keywords: user experience (UX) research, psychophysiological measures, agile development cycle, usability testing project, case study

1. Introduction

In many industrial fields, high-growth technologies are disrupting traditional models of development in multiple ways. On the one hand, expectations and demands of the market are changing at a faster pace, thus creating a competitive pressure for companies to launch their products at a higher rate than before. On the other hand, consumers, who are getting used to constantly seeking more, increasingly expect these technologies to fulfill their specific needs although the majority struggle to define what they really want and hope to have [1]. Hence, to meet the consumer's high expectations and needs, organizations are turning toward user experience (UX) research and its possibilities to put forth enriched measures that go beyond the deployment of explicit measurements resulting from interviews, focus groups, and questionnaires [2–4].

In order to better understand and get an overall picture of user interaction and satisfaction regarding a product, enriched UX measures arising from psychophysiological and neurophysiological data—that is, cognitive and emotional measures from lived experiences—have been proposed in recent research [5, 6]. However, we know from experience that analyzing such measures is time-consuming and complex; hence, they may not always be available promptly to inform product development [7, 8]. As a result, organizations find themselves in a methodological impasse: "There is no time to do thorough usability tests with users between iterations or release cycles, and only testing paper prototypes and doing expert analyses do not provide an accurate picture of the product's usability" [9]. They are thus expected to synchronize their production at a faster pace by adopting a rapid and efficient cycle of development while understanding the different aspects of the user's cognitive and emotional interaction.

It is, therefore, important to understand how to facilitate UX-enriched data collection and deployment by integrating an agile approach; a topic of growing interest that has been repeatedly projected in prior research presenting psychophysiological measures [10–12]. These implicit measures are less sensitive to social desirability and retrospective biases than explicit measures (e.g., self-reported questionnaires). Thus, the triangulation of explicit and implicit measures offers many advantages, such as providing richer and fewer biases in UX measures. This triangulation approach provides clarity on the participants' lived and perceived experiences [13–15].

This chapter investigates how these enriched measures can inform user experience evaluation while maintaining agility and speed in UX evaluation projects. Using a multiple case approach, we analyzed 12 recent usability testing projects in which enriched measures were used. It outlines the lessons learned with regard to challenges encountered, the advantages and limitations of using psychophysiological measures in UX evaluation, and the benefits for UX project management practice.

2. Literature review

The agile software development approach and the UX approach might appear to be conflicting, *a priori*, since they present two distinct ways of allocating resources within a project [16]. The two approaches are founded on different premises. The agile approach focuses on product development, while the UX approach stresses upon the harmonious integration of the object into the user's life—including the emotional engagement, hedonic appreciation, the values associated to the object, and the technological ecosystem in which the object is used [17]. Proposed iterations on well-defined functional sections of the project may not necessarily provide the same division of test units within the project [18]. The synchronization of activities and practices becomes complex. Indeed, the agile approach proposes a division of the project into working sets to be tested in interaction with the user to ensure their functionality. The UX approach, for its part, proposes a division of the project into needs of the user to be tested to ensure the quality of the specific and global experience of the user. Since there exists such a gap in achievement objectives of both these approaches, their integration will require good communication between all the stakeholders of the project as well as fine-tuning from the early stages of the project.

The agile approach thus provides a development structure to rapidly create products that fulfill the user's needs, while the UX approach provides the target user with a level of empathy, an element is lacking in the agile approach. In other words, on the one hand, the agile approach allows developers to create products that have value for the user: "Agile development lifecycle is characterized as a series of incremental mini-releases. Each mini-release, with a subset of the features for the whole release, has its own requirements analysis, design, implementation, and quality assurance phases, and is called a working version" [19]. On the other hand, the UX approach leads development teams to create products that are integrated harmoniously into the user's life and are adapted to them [20].

Nevertheless, both these approaches remain complementary: "Agile projects are highly feedback-driven, yet product teams often rely on user opinion in situations where observation is more appropriate (such as the focus group elicitation strategy described earlier)" [19]. Consequently, the UX approach can greatly improve the agile approach by providing a systematic and scientific way of assessing the needs of target users [18]. Yet, the integration of one approach within the other is complex since, within the UX practice, there are various types of measures implicating different time constrains. On the one hand, there are the neurophysiological data. The preparation needed to collect this kind of data is arduous but, with a strong methodology, can be analyzed rapidly. On the other hand, there is the perceptual data. This data are mainly collected through interviews and need a lot of time to analyze and assess. Finally, there is a promising avenue towards putting forward enriched UX measures implicating both of those data types [21].

There is a gap in the literature and a need to answer this crucial question: Can enriched UX measures be performed quickly enough to be include in an agile development? Two literature reviews on the subject [16, 18] present interesting conclusions and avenues of reflection. One of the main trends seems to be to promote a specialist approach through which the UX work within an agile team is carried out by a specialized designer researcher [22]. Collaboration and communication are also recurring themes in Agile/UX literature. Communication is highlighted not only by the application of the "scrum" model but also by the use of visual artifacts: "We find that both sketches and design stories have critical roles, that these artefacts support creation and reflection, facilitate resolution of contradiction, and also work at a level of consciousness that is below the level of self-awareness" [23]. In addition, to facilitate the integration of experiential results into UX within a development based primarily on product functionality, the Little Design Up Front (LDUF) practice is the most widely adopted initiative in Agile/UX [16]. "LDUF reduces—but does not eliminate—the large amount of design work done through [User-Centered Design] at the beginning of the project so that more effort can be spent on functionality" [18]. This practice is also enriched by the Sprint 0 (i.e., initial sprint), a Sprint process whereby initial user research is done so that all stakeholders can jointly create a basic skeleton and ensure that all future Sprints add incremental real value to the project.

UX designers also often have to simultaneously perform multiple roles involving numerous tasks such as user research, market research, user-centered design, prototyping, usability inspection, user testing, visual design, feedback, and coding [18]. Consequently, they are usually in different working groups, if not in several departments, or even different subcontracted organizations; this evidently complicates coordination and communication between all stakeholders. Moreover, UX researchers find themselves in a unique situation where they have to learn not only to adapt to a new culture and work environment but also to become quickly familiar with the project that has been granted to them. Often, a project may have already been initiated and it may even be in a phase of advanced development, thereby requiring UX researchers to work rapidly to take it forward. While immersed in an agile approach, UX researchers find themselves working on smaller sections of the project simultaneously instead of considering the whole project, which additionally tends to change fast.

With the purpose of integrating an agile approach into an UX research methodology, we must clearly define the objectives to be achieved and ensure that the expectations of all stakeholders are realistic and well defined from the outset of the project. Especially, since "Once there is an established relationship with the client, and the team is familiar with both how they work together and with outside resources, they can better assess the consultant's ability to work with them in an agile setting" [24].

The integration of the two approaches into a common methodology is based on two main strategies [25]: the first suggests that the UX team should become quickly integrated into the product development cycle so that it can understand the initial mission of the project and be present from the first decisions taken, and the second strategy suggests the use and deployment of "agile" tools to facilitate communication and documentation. These are mainly personas, usage scenarios, sketches, and concept maps to quickly understand the direction of the project as well as to facilitate message transmission to all the stakeholders of the project [23].

Regarding the importance of collaboration among the various stakeholders involved in the project, it is essential for all members to maintain constant communication and a working synergy to ensure the sharing of a common mission and vision. In addition, integrating targeted users at key points in the development process allows creators to respond appropriately to their needs. This way of working makes it possible to ensure a certain consistency and uniformity of the project as well as to more effectively control the expectations of the client. "In an ideal situation, UX development and research involves frequent, iterative user testing. Because agile focuses on smaller changes, it can be possible to conduct small-scale testing at various points throughout the process to ensure changes fit with UX expectations" [24]. These different parts of the project can also take the form of "Sprints" of the "development, testing, evaluation, and adjustment" cycle.

By adopting an agile approach, UX researchers tend to change their work methodology by reducing their activities, adopting a less formal process, and a more minimalist method [24]. Although the integration of an agile approach requires a restructuring of the UX experimental design, it is necessary to ensure that the integrity and enhanced value of the UX process are maintained, and even enriched with psychophysiological measures. The recent development of a laboratory management and analytics software platform for human-centered research now makes this kind of integrated process possible, which (a) enables accurate triangulation of enriched UX measures, (b) produces results in a timely manner, and (c) helps to generate meaningful recommendations [26].

3. Method

A multiple case study methodology was chosen as the preferred approach to investigate the project management practices that can be used to enable the enrichment of user experience evaluation while maintaining agility and speed in UX evaluation projects. We conducted 12 cases studies on usability test projects using enriched UX methods over a short period of time (maximum 2 weeks). All 12 tests were conducted by the same organization. In this chapter, we refer to these as a Sprint projects.

The multiple case studies thus make it possible to identify the inherent and recurrent markers [27–29] of the Sprint project's management practices in order to better define and understand it in all its complexity. Different variables of the Sprint projects have been highlighted to better understand its mechanisms such as the project's objective and its level of complexity, the execution (the UX team deployed and their work per hour ratio, the experimental design, the maturity of the stimuli (which were all prototypes), the tools used, the measures analyzed, and the time of completion), and, more specifically, the details of the tests (number of participants, recruitment process, and testing time which, for most of the projects, has been standardized to 12 participants and 1 h of testing). Finally, the degree of details in the test results has been presented in terms of the magnitude of the final report submitted (**Table 1**).

Data were collected using structured interviews with at least three members of each project. The structured interview covered the following properties for each project: (i) objective of the usability test; (ii) difficulty of the objective (1—easy to 5—difficult); (iii) description of the experimental design; (iv) maturity of the stimuli (prototype); (v) tools used and measures analyzed in the test; (vi) time to execute the test; (vii) difficulty of the execution (1—easy to 5—difficult); (viii) number of participants; (ix) population; (x) testing time (in minutes); and (xi) magnitude of the report (in pages). To evaluate the difficulty measures, we averaged the answer of the respondents. **Table 1** provides a summary description of all 12 projects that are presented in chronological order.

The interviews also included open-ended questions focused on project management practices. Questions covered project planning, project management, communication and coordination in the team, status of work with the external client, project execution, and analysis management.

4. Result

The 12 projects involved in this study are descripted in **Table 2**. In total, these projects necessitated the participation of 144 typical users (experts and neophytes), deployed 4 neurophysiological tools and 4 psychometric tools, and concluded with 799 pages of reports. It should be noted that the organization conducted regular debriefing sessions with the client to outline the failures and accomplishments. We can observe that over time, the projects experienced a significant reduction in execution time, human intervention, level of difficulty, and costs by standardizing the methodology. We went from a 19-day project to a 12-day project (including preparation time), from a 20 expert (internal staff and external sponsors) implication to a core team of only 4 experts, and from a level of difficulty of 5–2.5, which all ultimately affect the cost of operations.

Based on the interviews and the observation, it has been possible to put forward the following conclusions. To execute a Sprint projects, many considerations have to be taken into account:

1. The nature of the research: the nature of the research question impacts the completion time and the complexity of the project.

2. The nature of the elements:

    a. Human: need to communicate regularly with the design clients and various project stakeholders and jointly establish the mandate and experimental design with the concerned design clients.

    b. Technical: need to anticipate the collected measurements and enhanced results using a mosaic of hybrid collection methods.

3. The nature of the results: need to adapt the manner of presenting the results in order to underline the operational side without reducing the quality of the work performed.

Case	Objective			Execution			Details of the testing				Results
	Objective	**Level of difficulty (1—easy to 5—difficult)**	**Experimental design (condition = version of the product) (task = assignment)**	**Maturity of the stimuli (prototype or final product)**	**Measures***	**Time spent**	**Level of difficulty (1—easy to 5—difficult)**	**Number of participants**	**Sample**	**Testing time**	**Magnitude of the report**
1	Analysis of user behavior	5	3 conditions 3 tasks 1 interview 1 survey	Prototype	A&Cl *E* A (EDA) & A (EKG) KPI PI (SUS) N&I	14 days of preparation 2 days of data collection (testing) 3 days of analysis	5	12	Millennial –	1 hour	1 presentation 57-page report
2	Analysis of user experience of a mobile app	3	3 tasks 1 interview 3 surveys		A&Cl *E* A (EDA) KPI PI (SUS) N&I	7 days of preparation 3 days of data collection (testing) 2 days of analysis	3.5		Millennial –		1 presentation 41-page report
3	Analysis of user experience in interaction with a Chabot versus forms	3.5	2 conditions 4 tasks 1 interview 3 surveys		A&Cl *E* A (EDA) KPI PI (SUS) & PI (SAM) N&I	7 days of preparation 3 days of data collection (testing) 2 days of analysis	2.5		Millennial –		1 presentation 50-page report
4	Analysis of user experience in interaction with a transactional Website	2.5	4 tasks (3 sub-tasks) 1 interview 1 survey		A&Cl *E* A (EDA) KPI PI (Att) N&I	7 days of preparation 3 days of data collection (testing) 2 days of analysis	2		Millennial –		1 presentation 43-page report

Case	Objective		Execution					Details of the testing			Results
	Objective	**Level of difficulty (1—easy to 5—difficult)**	**Experimental design (condition = version of the product) (task = assignment)**	**Maturity of the stimuli (prototype or final product)**	**Measures***	**Time spent**	**Level of difficulty (1—easy to 5—difficult)**	**Number of participants**	**Sample**	**Testing time**	**Magnitude of the report**
5	Analysis of user experience in interaction with a transactional website, version 2.	3	4 tasks (4 sub-tasks) 1 interview 2 surveys	A&Cl *E* A (EDA) KPI PI (SUS) & PI (Att) N&I		7 days of preparation 3 days of data collection (testing) 2 days of analysis	2.5		Millennial –		1 presentation 35-page report
6	Analysis of user experience in interaction with a transactional website, version 3.	2.5	4 tasks (4 sub-tasks) 1 interview 2 surveys	A&Cl *E* A (EDA) KPI PI (SUS) & PI (Att) N&I		7 days of preparation 3 days of data collection (testing) 2 days of analysis	2.5		Millennial –		1 presentation 58-page report
7	Analysis of user experience in interaction with two different versions of a web and mobile interface	4.5	4 scenarios 15 tasks 2 conditions 4 interviews 1 survey	A&Cl *E* A (EDA) + A (EKG) KPI PI (Wq) N&I		7 days of preparation 4 days of data collection (testing) 4 days of analysis	4.5		Millennial –		1 presentation 102-page report

Case	Objective		Execution					Details of the testing			Results
	Objective	Level of difficulty (1—easy to 5—difficult)	Experimental design (condition = version of the product) (task = assignment)	Maturity of the stimuli (prototype or final product)	Measures*	Time spent	Level of difficulty (1—easy to 5—difficult)	Number of participants	Sample	Testing time	Magnitude of the report
8	Evaluate different age groups' user training and change management in interaction with a web site	2.5	6 tasks 1 interview 2 surveys	A&Cl *E* A (EDA) KPI PI (SUS) * PI (SAM) N&I		7 days of preparation 3 days of data collection (testing) 2 days of analysis	2.5		Millennial – & Baby boomer –		1 presentation 56-page report
9	Evaluate different age groups' user training and change management in interaction with a web site, version 2.	3.5	2 conditions 5 tasks per condition 1 interview 1 survey	A&Cl *E* A (EDA) KPI PI (Wq) N&I		7 days of preparation 3 days of data collection (testing) 2 days of analysis	3		Millennial –		1 presentation 54-page report
10	Analysis of user experience while opening an account on a smart phone mobile app	2.5	1 task 1 interview 3 surveys	A&Cl *E* A (EDA) KPI PI (SUS); PI (Wq) & PI (Att) N&I		7 days of preparation 3 days of data collection (testing) 2 days of analysis	2		Millennial –	30 minutes	1 presentation 49-page report

DOI: http://dx.doi.org/10.5772/intechopen.89762

Case	Objective		Execution				Details of the testing				Results
	Objective	Level of difficulty (1—easy to 5—difficult)	Experimental design (condition = version of the product) (task = assignment)	Maturity of the stimuli (prototype or final product)	Measures*	Time spent	Level of difficulty (1—easy to 5—difficult)	Number of participants	Sample	Testing time	Magnitude of the report
11	Analysis of user experience while opening a professional account on a smart phone mobile app	2.5	4 tasks 1 interview 3 surveys		A&Cl *E* A (EDA) KPI PI (SUS); PI (Wq) & PI (Att) N&I	7 days of preparation 3 days of data collection (testing) 2 days of analysis	2.5		Baby boomer –	1 hour	1 presentation 79-page report
12	Analysis of user experience during an online mortgage application process from a computer	3	1 task 1 interview 2 surveys		A&Cl *E* A (EDA) KPI PI (SUS) & PI (Wq) N&I	7 days of preparation 3 days of data collection (testing) 4 days of analysis	3.5		Gen X		1 presentation 75-page report

*Tools: *A&CL = attention and cognitive load is measured through Pupil and Gaze by a Tobii eye tracker; E = emotions are measured through facial expressions by FaceReader software; A = arousal is measured through electrodermal (EDA) and electroencephalogram (EKG) data with Biopac instruments; KPI = key performance indicators are measured by observation; PI = psychometric indices (PI) are measured through surveys such as the System Usability Scale (SUS), Webqual (Wq) and Attrakdiff (Att), and SAM scale (SAM); NI = customers need (N) and insights (I) are identified through interviews and analyzed with Optimal Workshop.*

Table 1.
Twelve case studies and their methodological insights.

		All projects	Mean and median per project
Execution	Experimental design	1–4 conditions 1–15 tasks 1–4 sub-tasks 1–4 interviews 1–3 questionnaires	2 conditions 4 tasks 2 sub-tasks 1 interview 2 questionnaires
	Tools used	3–4 neurophysiological tools 2–4 psychometric tools Observation (performance indicators)	2 neurophysiological tools 3 psychometric tools Observation (performance indicators)
	Time	7–14 days of preparation 2–3 days of data collection 2–3 days of analysis	7 days of preparation 3 days of data collection 2 days of analysis
Details of the experiment	Participants	144 typical users	12 participants
	Testing time	30 min to 1 h	1 h
Result	Final report	799 pages	66 pages

Table 2.
Statistics for the totality of the Sprint projects and mean per project.

4. The time constrains: (a) need to adjust the granularity (level of details) of the project according to the research question; (b) need to introduce pre-tests to provide last-minute adjustments on site; (c) need to carefully evaluate the time allotted for the project; and, thus, (d) need for scheduling.

These strategies are focused on meeting the clients' expectations of time, budget, and UX issues.

4.1 Based on the nature of the research

4.1.1 Research question type

Every UX research begins with a question. The nature of this issue has a direct impact not only on the completion of the UX tests but also on the complexity of the tests. This complexity depends on the nature of the stimuli studied and on the level of authenticity of the desired context of use. Indeed, the research question determines the nature of the stimuli, that is, whether they are static or dynamic. For example, studying the navigation of a Website on a computer screen underlies the deployment of static stimuli which, *a priori*, is easy data to analyze. Static stimuli require shorter coding and analysis time than dynamic stimuli, for example, the study of a game application on mobile. The same applies to the choice of data collection tools deployed. Coding and analyzing data from an eye tracker does not represent the same workload as coding and analyzing data from an electroencephalography (EEG) headset.

Moreover, the research question directly influences the choice of the context of use in which the experiment takes place and the importance of the level of authenticity to be respected. Inevitably, undertaking an experiment in a real-life context does not underlie the allocation and deployment of the same resources (material and human) and the same time space for its realization in a laboratory context. Dynamic stimuli and the context of authentic use are the most important limitations of Sprint projects. It is not said that they are not feasible in a short period of time, but the data that can be collected and the degree of analysis that can be achieved are more resource intensive. It is, therefore, important to explicitly communicate these limitations to the clients during the initial stages of development of the project, in order to limit the frustrations that they may generate. It is also important to note that this type of project cannot be applied in the fundamental research framework, although it relies on the results of this research to improve their structure. In other words, the co-researchers aim to propose project management structures that reflect the current industry needs.

4.2 Based on the nature of elements

4.2.1 Human: communication and coordination

Communication and coordination between different stakeholders of the project are key factors to the smooth functioning of the process. Communication is carried out by daily calls, and sometimes through meetings with the design clients or within the research team itself. The research team also invites various clients to attend the data collection to ensure that there is a common understanding of each step of the UX process. The quality of communication between the clients mainly influences the joint construction of the mandate and the experimental design. This step is vital for clarifying everyone's expectations as well as the potential results of the experiment.

> *"For example, in the 7th case study, three different clients were involved in the project. Consequently, our research team had to coordinate with all the clients to ensure that the understanding and expectations of the project were the same for everyone. In the final days leading up to the pre-tests, conference calls lasting from one to two hours with all the stakeholders were organized." (Project Manager)*

4.2.2 Technical: hybrid data collection method

The anticipation of measurements and results is also at the center of the agile/UX process developed by the research team. In parallel with the definition of the mandate and the division into use scenarios, the research team continually tries to foresee the structure of the presentation of the results while being flexible. Empirical data, both implicit (lived experience assessed with psychophysiological measures) and explicit (perceived experience assessed with self-reported questionnaire and interview), are considered. This anticipation is carried out using a systematic methodology of foreseen codification of the psychophysiological—emotional and cognitive—measures within the clarification of the mandate and the experimental design. The triangulation of measures also makes it possible to anticipate the potentially interesting results that will answer the client's questions. This triangulation is achieved through a mosaic of proven collection methods [5, 30, 31]. The use of several data collection technologies of variable nature (physiological, psychological, and behavioral) ensures an enriched data collection. Consequently, this anticipatory effort allows the UX team to be one development cycle ahead of others and to accelerate the whole process of analyzing the collected data. Comparative empirical data methodology is also deployed. By comparing different conditions of use, design elements, or even groups of users, decision-making becomes more objective, concrete, and easy for the team of designers.

> *"For example, in the 7th case study, the project involved the collection of implicit data from eye tracking (Tobii), recognition of facial expressions (Facereader,*

Noldus), electrodermal activity (Biopac) as well as electrocardiogram (Biopac) and explicit data from usability scale questionnaires, performance indicators and interviews. This arsenal of tools was deployed to understand the "what and when" of interaction by triangulating valence (positive or negative), activation (weak or strong) and cognitive (easy or difficult) reactions, as well as the why of interaction through the verbalization of perceived experiences. The hybridization of all the data on the cognitive and emotional load thus created a global portrait of the interactive experience between the users and the product." (UX Lead)

4.3 Based on results

4.3.1 Data visualization

Finally, the lab team has developed a unique and innovative way of presenting its results to facilitate the transmission of knowledge to clients and development teams. By aggregating and triangulating the arsenal of empirical data collected, the laboratory's researchers have succeeded in creating a methodology for simplifying and making the data more accessible. The results of this methodology are the visualizations of the interactions through the creation of UX heatmaps [5, 30, 31]. These heatmaps offer an "easy to interpret UX evaluation tool which contextualizes users' signals while interacting with a system. Using these signals to infer the users' emotional and cognitive states and mapping these states on the interface provide researchers and practitioners with a useful tool to contextualize users' reactions" [10].

"For example, in the 8th study case, the presentation of the final report including the results of the UX research was carried out with the client's design team and several decision-makers. Using empirical and perceptual data visualization tools, managers from different departments who do not face this type of research on a daily basis quickly realized which of the products studied best met the usability objectives, thus having clear facts with which to make their decision." (UX Lead)

4.4 Based on time constrains

4.4.1 Granularity

The granularity of the project follows the definition of the mandate. Generally, the research team uses a list of questions from the clients as a baseline to translate them into defined actions. In other words, the UX team restructures the project by dividing it into different evaluation conditions. These conditions typically result in distinct usage scenarios that are not necessarily related to product functionality. These similar condition divisions allow the UX team to define the evaluation markers, as well as the performance indicators more easily, in order to facilitate the assessment of the overall and specific user experience.

"For example, in the 2nd case study, the customer wished to evaluate the efficiency and efficacy of three functionalities of its new product in development. After numerous exchanges, our research team translated this mandate into an operational experimental design that included the testing of both their old and new products with two different comparable evaluation conditions. The first one consisted in testing the 3 functionalities on the old product with existing users in order to establish a comparison baseline. Then, by deploying the theory of learning, the three functionalities were tested randomly three times on the new product. The third repetition was the one that was compared between products." (UX Lead)

4.4.2 Pre-test

As each project has its own specificities and distinguishes itself from others, pre-tests are always necessary. Undertaken in a short time span, these pre-tests allow the UX team to make final adjustments before starting the data collection with the participants. Three pre-tests are usually performed. The first is a technical test to ensure that all collection and analysis instruments are functional and set up properly to facilitate collection. The second is done with a member of the team to evaluate the time and fluidity of the experimental task. The third test is done with an external participant to ensure the understanding of each step of the experimental task and to avoid any misunderstanding during the data collection.

> *"For example, in the 6th case study, while performing the pre-tests, our research team realized that one of the tasks could not be done in the sequence that was proposed initially, and this caused a major change in the experimental design and protocol. The pre-tests prevented loss of data from one of the recruited participants." (Lab Manager)*

4.4.3 Standardization of the planning and methodology

With each successive Sprint project, the research team gradually standardized the process and practice to enhance their execution in terms of speed, efficiency of human resources, and costs. Indeed, the team put together a concise timetable detailing every step of a Sprint project where responsibilities for the research team and the design client are granted, and deadlines are specified. This timeline presents, on one hand, the elements of macro-planning in terms of weeks. Depending of the maturity and knowledge of the design client about their context of intervention as the product or service they wish to test, this preparation phase is variable and flexible. Furthermore, as regards the academic context, the submission of the ethics certification (considering the academic research context) requires many weeks of anticipation, since this is to ensure that all approvals have been obtained before starting the user experience testing. However, if the Sprint project is a sequel to a previous one or if a design client has already made a Sprint project and wishes to carry out a second one, this preparatory phase gradually decreases in terms of time since it increases in terms of efficiency. On the other hand, the elements of micro-planning in terms of days and hours, such as details of the execution, are specified and are the main interest of this standardized timeline (**Table 3**).

This normalized timeline presents the critical path of a Sprint project: (1) project kick off; (2) mandate definition; (3) experimental design fine tuning; (4) pre-test and validation; (5) data collection; (6) codification; (7) analysis; and (8) final presentation. Aiming to be completely transparent, this normalized timeline's intentions are to help all the project stakeholders to understand the critical steps that could delay the project, identify the persons in charge of the various steps, so as to avoid any misunderstanding and repetition of efforts. Moreover, it can be taken as a list of actions to be considered when starting a UX research project.

4.4.4 Time allotted

Another important aspect to consider during Sprint projects is the time allotted for carrying out the tests. This similar aspect turns into one of the limitations of agile/UX research. Indeed, for a Sprint project to be realized in 1 week, the experience of using the evaluated product or service can hardly exceed 1 h without having a direct consequence on the realization and costs. The time allotted for data

	Responsibilities					Sprint										
		M-1	D-14	D-7	D-5	D-3		D-2	D-2	D-1	D1	D2	D3	D4	D5	D7
						09:00	10:00	10:00	14:00	14:00						
Initiation of the project																
Initial meeting	UX team/client															
Ethics certification submission (according to the context of intervention: academical vs. industrial)	UX team															
Contract	Client															
Clarification of the mandate																
Internal planning (room reservation, etc.)	UX team															
Compensation for the participants	UX team/client															
Client's involvement	Client															
Hours of data collection	UX team/client															
Definition of roles and responsibilities	UX team/client															
Final mandate	Client															
Recruitment criteria	UX team/client															
Fine tuning of the experimental design																
Experimental design	UX team															
Delivery of the first version prototype	Client															
Questionnaires	UX team															
Recruitment	UX team or client															
Validation of the experimental design	UX team/client															
Number of markers	UX team															

	Responsibilities	Sprint													
		M-1	D-14	D-7	D-5	D-3	D-2	D-2	D-1	D1	D2	D3	D4	D5	D7
Planning of the markers	UX team														
Pre-tests and validation															
Final prototype delivery	Client														
Prototype validation	UX team														
Technical pre-test	UX team														
Participants' list	Client														
Compensation	UX team or client														
Last minute adjustments on the prototype	Client														
Internal pre-test	UX team														
Internal validation of the experimental design	UX team														
External pre-test	UX team														
External validation of the experimental design	UX team														
Protocol adjustment	UX team														
Data collection															
Day 1 of data collection	UX team														
Day 2 of data collection	UX team														
Codification															
Extraction	UX team														
Codification	UX team														
Analysis															
Analysis	UX team														

	Responsibilities														**Sprint**
		M-1	**D-14**	**D-7**	**D-5**	**D-3**	**D-2**	**D-2**	**D-1**	**D1**	**D2**	**D3**	**D4**	**D5**	**D7**
Report preparation	UX team														
Presentation															
Report presentation	UX team														

Table 3.
Macro and micro planning standardization.

collection shall inevitably include: (a) greeting of the participant; (b) signature of the consent letter; (c) assembly and calibration of the data collection apparatus; (d) performance of the experimental tasks; (e) questionnaires and interviews; (f) removal of the equipment; and (g) handing over of the compensation. Consequently, the completion of the three series of pre-tests takes on added importance, as it allows the UX team to ensure that the time allocated is not exceeded.

4.4.5 Scheduling

Finally, the third major challenge is to meet the strict timetable laid out by every project client. Delays in defining the protocol or delivering the ready-to-test prototype have a direct impact on the implementation of the Sprint project. Depending on the availability of experimental rooms, delays may postpone the project for several days, weeks, and even months. They can also generate significant costs for each of the project clients.

5. Discussion

In a *Sprint project*, the deployment of these strategies poses many challenges for the UX team, which is responsible for clarifying the nature of the questions asked and the time allotted for carrying out the tests to ensure compliance with the timetable. These challenges inevitably compromise the balance of the iron triangle—scope, time, and cost—[32] defined in project management as the key to quality.

As mentioned before, each of the *Sprints projects* ended with a debriefing between our research team and the client(s) involved in the project. These debriefings shed light on the mistakes made and the improvements for the subsequent projects on which iterations were made. From these improvements, it is possible to cite: (1) the level of management on the project which directly points towards; (2) improved communication in the preparatory phase; (3) the setting up of a statement of work; and (4) the training of research assistants.

5.1 Leadership, management, and communication

In a context where we are in an academic research lab doing applied research with industrial clients, those last ones are not always educated about the possibilities and limitations of UX testing that can be carried out. It is, therefore, important for the main research team to educate and guide clients when it comes to defining the research question and experimental design. First-time projects with new clients inevitably require additional time and effort in the preparatory phase. Bearing this in mind, the research team must, therefore, develop an educational strategy to facilitate this phase.

5.2 Statement of work

With the same objective of facilitating communication and building a common project, the establishment of a statement of work (SOW) can be considered. It may not always be possible to fulfill every desire of the client in a laboratory setting. It is, therefore, important to set ground rules that clearly distinguish which aspects of the project are flexible from those that are less adaptable, in order to limit subsequent frustrations on both sides of the stakeholders.

5.3 Training of research assistants

Finally, it is also important to consider the academic setting in which the lab team operates. Consequently, research assistants are students so that we must continually trained and mentored. This creates an extra level of preparation. UX tests require, *a priori*, rigorous preparation. Since research assistants are very involved in collecting and codifying the collected data, they must be able to understand everything that these UX tests imply.

Hence, an examination of the different case studies of *Sprint projects* has allowed us to highlight several strategies and lessons learned in the hybridization of an agile/UX approach. Notwithstanding, far from being a definitive methodological proposal that meets all the requirements of an efficient approach, the co-authors especially wanted to shed light on interesting lines of thought. The objective of such research is to find an approach that maintains and makes the iron triangle of project management more sustainable, i.e., reducing the time and operational costs while maintaining the quality and scope required by the client. Research is, therefore, continuing in this direction not only in improving the operationalization of such methods but also to make progress in the efforts to systematize the codification, analysis, and visualization of enriched UX measurements.

Since it is a new way of thinking Agile and UX methodologies, it is difficult to find different organizations that implement this methodology, explaining why all case studies are principally conducted within the same laboratory. Even though it helps to test and improve the methodology, it is also considered as a limitation. Since there are clear benefits for industrial practice in this area of research, studying other organizations that implement a similar or different methodology—merging Agile and UX approaches—would be beneficial for in depth and future research.

On the one hand, the classic UX research model that generally prioritizes the perceptual facet of UX (focus group, interview, questionnaire, and observation) is enriched by psychophysiological UX measures. On the other hand, given the costs involved in purchasing the devices, their deployment, and the training required for the application of psychophysiological-enriched UX measures, the co-authors sees a great opportunity for complementarity and transfer of knowledge between industry and academia, principally because it allows the industry to take advantage of the academic environment and research context, in order to explore the best avenues for an agile/UX development approach.

6. Conclusion

In conclusion, to the initial question as to whether enriched UX measures can be performed quickly enough to be included in an agile development, the answer is, therefore, yes. To create a UX project that follows the agile development guidelines, the key to success is to be able to steer its approach on numerous small incremental phases oriented on the users. The working sets as defined in the agile development oriented in functionalities should, as far as possible, be aligned with the users' needs and be tested by them early in the process, and later on during all the development phases, to insure that the final product is integrated harmoniously into the user's life and is adapted to his or her needs.

Furthermore, as proved by the case studies, significant findings could be made from testing the concepts and prototypes on participants. Therefore, it is important not to wait to have a finished and polished product before involving the user and having his or her perspective. With as few as a dozen participants and with a timeline of 2 weeks, it is possible to obtain quick insights that redirect the project and that better align it with the real user's needs. The tests could be made on specific features, as well as on a complete product aiming to quickly eliminate erroneous assumptions. According to what is tested, different tools are available and various approaches can be deployed. It is then that the UX designer's expertise becomes important to identify which tools and approaches will help to obtain the desired answers.

However, as easy it may seem at first glance, the variables to be considered are numerous and deeply exposed in this article—the nature of the research, the nature of the elements (human and technical), the nature of the results, and the time constrains. Several answers about how the research team has been able to improve their effectiveness can be found in the article. Nonetheless, adding to these variables, it is also important to consider the prevalent mentality in certain industries, and very well anchored in conservative industries—such as banks, insurance, governments, etc.—that often do not want to test concepts or ideas—based on a fear of industrial espionage or reputation issues—and wait until they develop a full and finished product. Within these specific industries, an educational phase will be the first step to implement innovative approaches that aim towards agility and speed in enriches UX evaluation projects.

Author details

Juliana Alvarez[1*], David Brieugne[2], Pierre-Majorique Léger[2], Sylvain Sénécal[2] and Marc Frédette[2]

1 Montreal, Canada

2 Tech3lab, HEC Montréal, Canada

*Address all correspondence to: juliana.alvarez.dehaut@gmail.com

References

[1] Cohen D, Lindvall M, Costa P. Agile software development. DACS SOAR Report 11. Fraunhofer Center Maryland, USA; 2003

[2] Cooper A, Reimann R, Cronin D. About Face 3: The Essentials of Interaction Design. New York City, United States: John Wiley & Sons; 2007

[3] Goodwin K. Designing for the Digital Age: How to Create Human-Centered Products and Services. John Wiley & Sons; 2011

[4] Hartson R, Pyla PS. The UX Book: Process and Guidelines for Ensuring a Quality User Experience. Elsevier; 2012

[5] Courtemanche F, Léger P-M, Dufresne A, Fredette M, Labonté-LeMoyne É, Sénécal S. Physiological heatmaps: A tool for visualizing users' emotional reactions. Multimedia Tools and Applications. 2018;77(9):11547-11574

[6] Georges V, Courtemanche F, Sénécal S, Léger P-M, Nacke L, Pourchon R. The adoption of physiological measures as an evaluation tool in UX. In: Proceedings Paper Presented at the International Conference on HCI in Business, Government, and Organizations. Vol. 10294. Vancouver, BC, Canada: Springer; 9-14 July 2017

[7] Law ELC, van Schaik P, Roto V. Attitudes towards user experience (UX) measurement. International Journal of Human-Computer Studies. 2014;**72**(6): 526-541

[8] Vermeeren AP, Law ELC, Roto V, Obrist M, Hoonhout J, Väänänen-Vainio-Mattila K. User experience evaluation methods: Current state and development needs. In: Proceedings of the 6th Nordic Conference on Human-Computer Interaction: Extending Boundaries. ACM; 2010. pp. 521-530

[9] Sohaib O, Khan K. Integrating usability engineering and agile software development: A literature review. In: Paper Presented at the Computer Design and Applications (ICCDA), 2010 International Conference. Qinhuangdao, China; 25-27 June 2010. p. 38

[10] Georges V, Courtemanche F, Senecal S, Baccino T, Fredette M, Leger P-M. UX heatmaps: Mapping user experience on visual interfaces. In: Paper Presented at the Proceedings of the 2016 CHI Conference on Human Factors in Computing Systems. California, USA; 2016

[11] Georges V, Courtemanche F, Sénécal S, Baccino T, Léger P-M, Frédette M. Measuring visual complexity using neurophysiological data. In: Information Systems and Neuroscience. Springer; 2015. pp. 207-212

[12] Léger P-M, Sénécal S, Auger C. Le défi de l'expérienceutilisateur. Gestion. 2015;**40**(2):50-57

[13] De Guinea AO, Titah R, Léger P-M. Measure for measure: A two study multi-trait multi-method investigation of construct validity in IS research. Computers in Human Behavior. 2013; **29**(3):833-844

[14] De Guinea AO, Titah R, Léger P-M. Explicit and implicit antecedents of users' behavioral beliefs in information systems: A neuropsychological investigation. Journal of Management Information Systems. 2014;**30**(4): 179-210

[15] Tams S, Hill K, de Guinea AO, Thatcher J, Grover V. NeuroIS-alternative or complement to existing

methods? Illustrating the holistic effects of neuroscience and self-reported data in the context of technostressresearch. Journal of the Association for Information Systems. 2014;**15**(10):723

[16] Da Silva TS, Martin A, Maurer F, Silveira M. User-centered design and agile methods: A systematic review. In: Paper Presented at the Agile Conference (AGILE). Salt Lake City, USA; 2011. p. 2011

[17] Robert J-M. Vers la plénitude de l'expérienceutilisateur. In: Paper Presented at the Proceedings of the 20th Conference on l'Interaction Homme-Machine. New York, USA: 2008. p. 4

[18] Jurca G, Hellmann TD, Maurer F. Integrating agile and user-centered design: A systematic mapping and review of evaluation and validation studies of agile-UX. In: Paper Presented at the Agile Conference (AGILE). Vol. 2014. 2014. p. 25

[19] Sy D. Adapting usability investigations for agile user-centered design. Journal of Usability Studies. 2007;**2**(3):113

[20] Ferreira J, Sharp H, Robinson H. User experience design and agile development: Managing cooperation through articulation work. Software: Practice and Experience. 2011;**41**(9): 963-974

[21] Krawczyk P, Topolewski M, Pallot M. Towards a reliable and valid mixed methods instrument in user eXperience studies. In: Engineering, Technology and Innovation (ICE/ ITMC), 2017 International Conference on. Madeira Island, Portugal: IEEE; 2017. pp. 1455-1464

[22] Fox D, Sillito J, Maurer F. Agile methods and user-centered design: How these two methodologies are being successfully integrated in industry. In: Paper Presented at the Agile, 2008. AGILE'08. Conference. Toronto, Canada; 2008. pp. 63-72

[23] Brown J, Lindgaard G, Biddle R. Stories, sketches, and lists: Developers and interaction designers interacting through artefacts. In: Paper Presented at the Agile, 2008. AGILE'08. Conference. Toronto, Canada; 2008. p. 39

[24] Meingast M, Ballew T, Edwards R, Nordquist E, Sader C, Smith D. Agile and UX: The road to integration the challenges of the UX practitioner in an agile environment. Paper Presented at the Proceedings of the Human Factors and Ergonomics Society Annual Meeting. San Diego, California, USA; 30 September, 4 October 2013. p. 1002

[25] Kollmann J, Sharp H, Blandford A. The importance of identity and vision to user experience designers on agile projects. In: Paper Presented at the Agile Conference, 2009. AGILE'09. Chicago, USA; 2009. p. 12

[26] Léger P-M, Courtemanche F, Fredette M, Sénécal S. A cloud-based lab management and analytics software for triangulated human-centered research. In: Proceedings NeuroIS Retreat 2018. 2018

[27] Baxter P, Jack S. Qualitative case study methodology: Study design and implementation for novice researchers. The Qualitative Report. 2008;**13**(4): 544-559

[28] Tellis WM. Application of a case study methodology. The Qualitative Report. 1997;**3**(3):1-19

[29] Yin RK. Case Study Research: Design and Methods. Sage Publications; 2013

[30] Léger P-M, Davis FD, Cronan TP, Perret J. Neurophysiological correlates of cognitive absorption in an enactive training context. Computers in Human Behavior. 2014;**34**:273-283

[31] Léger P-M, Sénécal S, Courtemanche F, de Guinea AO, Titah R, Fredette M, et al. Precision is in the eye of the beholder: Application of eye fixation-related potentials to information systems research. Journal of the Association for Information Systems. 2014;**15**(10):651

[32] Atkinson R. Project management: Cost, time and quality, two best guesses and a phenomenon, it's time to accept other success criteria. International Journal of Project Management. 1999; **17**(6):337-342

Chapter 7

Collaborative Production Model of Educational Resources for Human-Computer Interaction Community in Latino America

Jaime Muñoz-Arteaga

Abstract

Today, human-computer interaction (HCI) shows great activity, dynamism, and academic presence throughout the world, having special relevance in the Latin American region. The region is not only vast for its geographical space but also diverse and multicultural, where researchers and academics from this area or community have proposed and made known the benefits of the HCI that they can bring to today's society. However, some problems arise and need to be addressed in the HCI area in the Latin American region, such as the lack of training strategies and the availability of content and educational resources in Spanish. In order to mitigate this problem, the current work proposes a collaborative production model of educative resources for human-computer interaction developed in Latin America. The model preconizes a series of strategies and technological services to support the collaborative production and access of HCI educative resources such as videos, slides, handouts, textbooks, user experience analysis, and usability tests. The proposed model is tested throughout two real case studies conducted by teachers and researchers from different Latin American universities in order to produce and use the HCI educative resources for under- and postgraduate courses.

Keywords: HCI, webinars, textbooks, educational resources, collaborative production, SOA

1. Introduction

Nowadays, the area of research and education specializing in human-computer interaction (HCI) shows a high activity and a dynamic presence among academic groups from universities and research centers around the world. Human-computer interaction is an area of computer science that studies human interaction with computers. In other words, HCI studies how to design, develop, and evaluate new computational technology and information, in such a way they become easy to use and useful to human activities [1]. HCI plays a major role in the design of interactive systems since the nature of its knowledge body is multidisciplinary, for example, software engineering, sociology, computer science, neuroscience, design, artificial intelligence, and cognitive psychology [1–3].

In the last decade, the HCI has a special relevance in the Latin American region. A large number of researchers of HCI community have presented their contributions at several conferences [4]. However, some problems arise and need to be addressed in the HCI area in the Latin American region, such as the lack of training strategies and the availability of content and educational resources in Spanish. One solution is to develop HCI contents in terms of different formats such as demos, slides, handouts, textbooks, user experience analysis, and usability tests. The advantage of digital educational resources versus traditional formats is that they can be accessed and reused to be part of new online courses with different learning styles at a university level. The user can have access to educational resources with additional multimedia features (such as video, audio, animation and interactive applications, and 3D content), and finally they can update as often as necessary [5].

This work presents a model as an alternative solution for the collaborative production of educational resources, emphasizing various types of online resources available to the user. It is structured as follows: the next section shows the production issue of HCI education resources in the Latin American region; Section 3 presents a conceptual model for the collaborative production of education resources in HCI. The two subsequent sections implement the model in two real case studies where teachers and researchers from several universities cooperate in order to provide educational resources available in Spanish to the HCI community of Latin America.

2. Problem outline

The region of Latin America is not only vast for its geographical space but also diverse and multicultural, where researchers and academics from this area or community have proposed and made known the benefits of the HCI that they can bring to today's society.

Unfortunately, in Latin America, it has been more difficult to use the traditional educational resources, largely due to high costs and the lack of access to services. In addition, the publications that are in a foreign language with different cultural and educational contexts make this type of materials inaccessible for students at the Institution of Higher Education (IHE) in Latin America [6]. One reason for the high cost of textbooks is that most of them are produced outside the region [4]. The fundamental problem is not related to the lack of production capacity but to the difficulty of teachers or local authors to publish and distribute their own educational resources. The problem with the origin of the textbooks has several additional consequences, besides their cost [7]: most educational resources are not adapted to the context of higher education in Latin America, the latest versions being not available in a language mastered by most teachers and students, which creates a damaging perception of being in a lower position and that knowledge always comes from abroad [8].

In general, some educational resources produced in Spain are used by teachers since the beginning of this century, but the contents require explaining and adapting to Latin American educative context. One should not forget that one of the biggest problems is the lack of books, educational content, and related resources in this area [5, 6]. Other problems arise and need to be addressed in the HCI area in the Latin American region, such as the lack of training strategies and the availability of content and educational resources in Spanish as shown in **Figure 1**. According to that figure, note that it is also necessary to consolidate the technical skills in the area and establish communication mechanisms between professionals and academics in the future development of the HCI. The problem of access to HCI content for both teachers and students at a university level has several additional consequences, apart from its cost [9]; most of the contents are not adapted to the Latin American

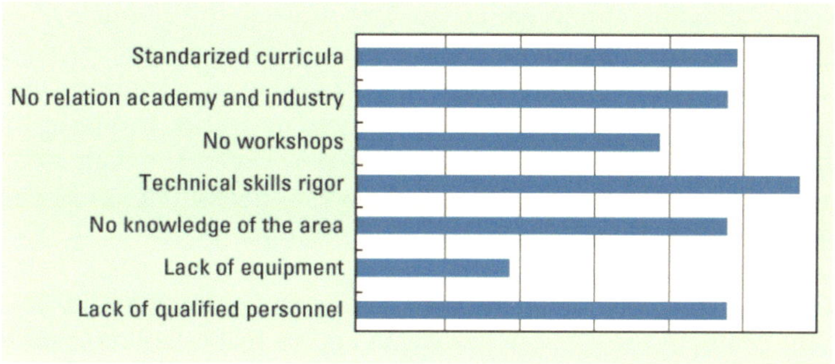

Figure 1.
Some issues related to teaching human-computer interaction [4].

context, and the most recent versions are not available in a language most teachers and students dominate, creating a not very good learning experience and perceiving knowledge in general comes from the outside.

3. Collaborative production model of educational resources for HCI

This section proposes an extension of spiral model from Boehm software process model [10] to describe a collaborative production of HCI educational resources (see model of **Figure 2**). This model is characterized as an iterative, incremental, and interactive process. Then, in the early stage of planning phase, it is necessary to define a plan for the production of new HCI educational resources as well as the organization of human resources. It is involving here system analysts, media specialists and content producers are involved here. The next phase (design phase)

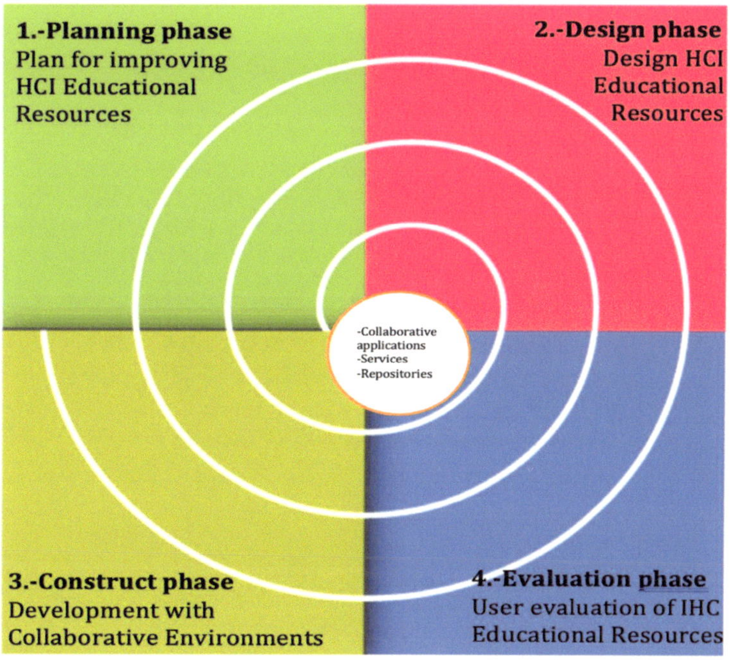

Figure 2.
Collaborative production model of HCI educational resources (inspired of Boehm's model).

focuses on the validation of new educational resources at the design level before launching a massive development of these resources. The construct phase is for the development of educational resources using collaborative environment; the process is directly influenced by learning and social networks, involving practice communities and collaborative tools. The fourth phase is for the evaluation of HCI educational resources using repositories to allow the content that is retrieved, recommended, appointed, grouped, and so on to be a contender to the future versions of the course.

Of note the collaborative applications, services, and online repositories are located in the center of the proposed model in **Figure 1**; at least one of these key elements is necessarily used by the participants in every phase of the development of educational resources. In fact, a multidisciplinary team is required to implement the previous model, including teachers, reviewers, researchers, content organizer, and content producer. They are involved in a huge effort of creating resources that are designed for use in formal and non-formal educational situations. The whole responsibility lies with the multidisciplinary team to provide creative content and give certain guarantee that it is the most suitable content that could be offered to each specific learning/teaching situation [5].

The next two subsequent sections implement the previous model in two real case studies where teachers and researchers from several universities cooperate in order to produce educational resources in Spanish language to the HCI community of Latin America.

4. Collaborative production of textbooks

The Latin project has developed an environment for the collaboration and production of a series of textbook in computer science such as the HCI book (**Figure 3**), and then the teachers can contribute whit sections or chapters to be put together into custom textbooks for the whole community.

Title: Temas de Diseño en Interacción Humano-Computadora

Categories: Ingeniería

Authors: Jaime Muñoz Arteaga, Yosly Caridad Hernández, Viviana Bustos Amador, Anita Aranda Chavarría, Marta E. Calderón, César E. Collazos, Yenny Alexandra Méndez, Andrés Fernando Solano, José Eder Guzmán Mendoza, Francisco Álvarez Rodríguez, Ricardo Mendoza González, Josefina Guerrero García, Juan Manuel Gonźales Calleros, Liliana Rodríguez Vizzuett, Toni Granollers, Rosa Gil, David Céspedes Hernández

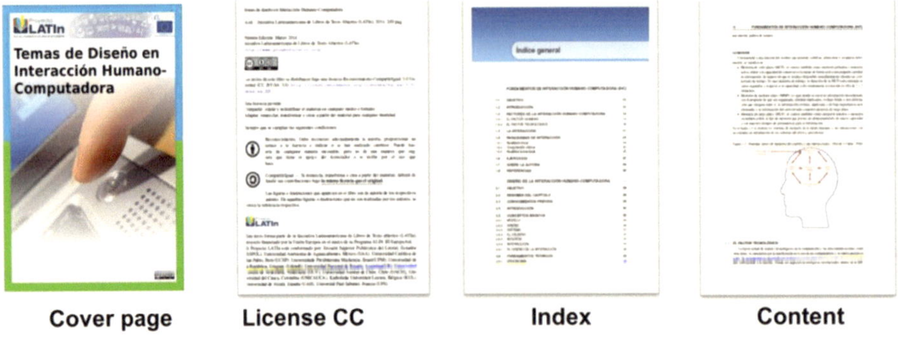

Figure 3.
Excerpt of Chapter 1 from the HCI textbook [11].

The previous figure shows an excerpt from the HCI textbook titled *Temas de Diseño en Interacción Humano-Computadora*, in which the index and the whole content of this book is written in Spanish language with creative common license. All the chapters have a similar structure, namely, the objective of the document with theoretical and practical sections with solved exercises, self-evaluation, and bibliography [4]. The solved exercises illustrate how to apply the theoretical knowledge, and the evaluation sections allow the future readers a self-evaluation of HCI subjects.

4.1 Collaborative applications

The writing textbook process starts with the formation of teams of collaborative work among the researcher teachers related with their knowledge area. The Latin project has built an application for collaboration and production of the HCI book, so that teachers and authors can contribute whit sections or chapters to be assembled into custom books for the whole community.

Figure 4 shows several professors and researchers of different HCI; they were members of an HCI writing group, thanks to the collaborative tools available. In addition, we defined a first title about the proposal and some ideas as a part of first process iterations, being consisted in whole writing group information's HCI textbook.

On the other hand, **Figure 5** shows the color-coded roles of the actors: the general manager is represented in brown, the communicator in green, the idea generator in purple, etc. The different parts of this structure can be assigned to one or more authors as a coordination process for writing. Author "A" can work or comment anywhere in the document, while author "B" is allowed to read just any part of a chapter, and author "C" cannot see that the rest of the document has not been assigned. However, the teachers participating in the book of HCI held different roles. **Figure 5** also illustrates the distribution of set chapters of HCI book as a result of interactions and collaborative activities of authors.

Then, the specific process to produce the HCI textbook is composed of four phases according to the proposed model of **Figure 2**. In the first phase a call for participation was proposed for the collaborative teamwork among HCI teachers of several higher education institutions who want to have educational resources such

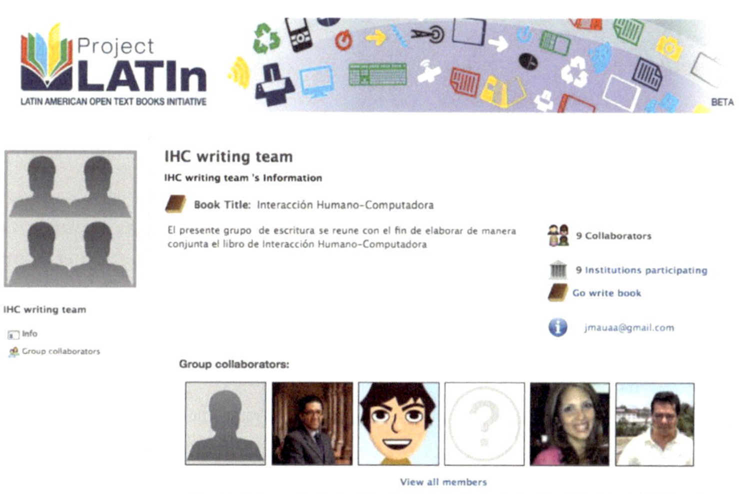

Figure 4.
HCI writing team in the Latin platform.

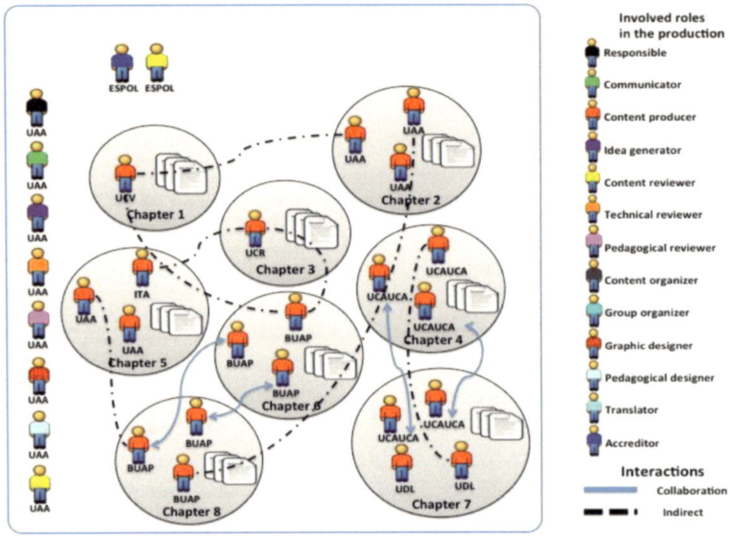

Figure 5.
Diagram of collaborative production by teachers from several universities of Latin America.

as textbooks. The second phase starts with the reception of book proposals by the defined collaborative work groups; these groups through social interactions define the theme on which they will create their contents and make contributions on ideas that will be useful to them through the use of chats, forums, messages, discussion groups, etc. The third phase is the production process of open textbooks. Once the collaborative team has the green light to create a proposed book, then it is possible to use of synchronous and asynchronous tools and activate the functionalities that provide collaboration services for an appropriate work, for example, text editors, review templates, and communication tools. Examples of these last tools are chats, mails, blogs, forums, resource sharing, ideas, etc.

The evaluation of open textbooks is the fourth phase. Once the textbook is complete, it will be submitted to an evaluation process where a technical evaluation is carried out. The reviewers and members committee have used this platform to make technical evaluation about the proposed books. After the technical evaluation, a resolution of acceptance or rejection of the proposed textbook is issued. It recognizes also the authorship of the collaborative team members. Finally, the open textbook is released under an online version.

4.2 Services

The technology of service-oriented architecture [12, 13] was exploited to identify several web services to support the main process of the collaborative production of an HCI textbook such as collaborative textbook production and textbook management (see **Figure 6**). The web services for the first process are as follows: defining textbook chapters, collaborating textbook design, planning meeting, and collaborative edition of textbooks. The second process of textbook management is supported by the web services: call for a new textbook, define collaborative team, identify authors, identify reviewers, evaluate textbook proposal, review textbook versions, and release new textbook.

In this case study, the collaborative production of educational resources is concreted in the repository of HCI textbook from the Latin project. In addition, the participants in the HCI webinars become members of the LACLO community [14].

DOI: http://dx.doi.org/10.5772/intechopen.90992

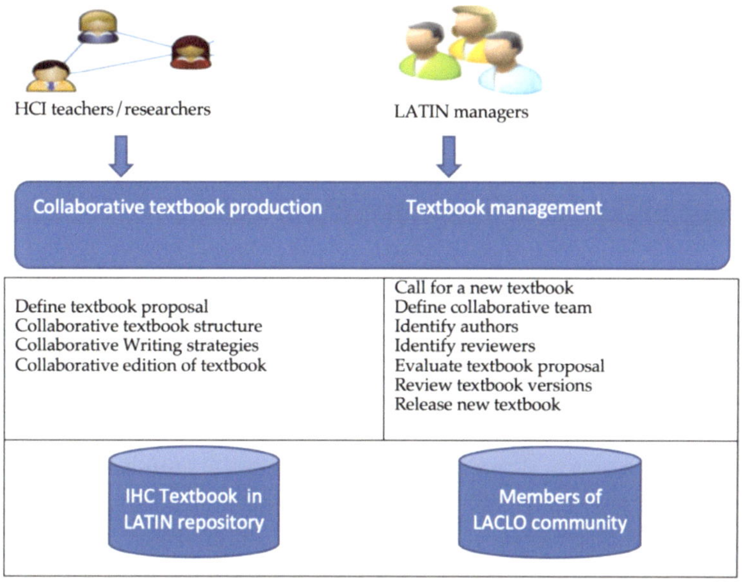

Figure 6.
Web services for collaborative production of HCI textbook.

4.3 Repository

A repository for textbook was built where the HCI book is located. The HCI textbook [11] is composed of a set of important chapters such as the foundations and specialized topics in HCI, including chapters about design, development, and evaluation of interactive applications (see **Figure 7**). The book can be used as a basic support of a study material for HCI courses at both undergraduate and graduate levels.

The structure of the textbook consists of several elements that are common among the publishing features of a printed book, such as the cover, foreword, preface, introduction, etc. The different parts of this structure can be assigned to one or more authors as a coordination.

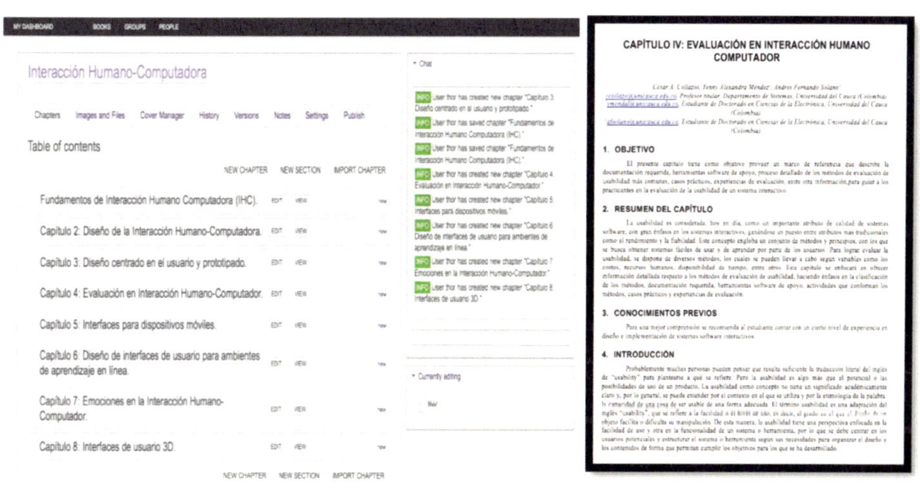

Figure 7.
List of chapters from the HCI textbook [11].

5. Collaborative production of HCI web seminar

Another strategy to expand the dissemination of HCI in the Latin American region is to invite academics, professionals, and researchers from HCI to publish their research topics through a webinar ("web" more "seminar"), where they will be able to present and share their experiences, projects, and research in HCI to direct a large number of stakeholders from students, entrepreneurs, and academics and researchers themselves [15].

Among different social media technologies, the webinar is increasingly recognized and used throughout the world; in the context of Latin America, it has as one of its main advantages the possibility of effective communication between people from different geographical places. This facilitates the attendance at conferences and events from anywhere in the world, both in real time and also in deferred. For example, the HCI webinars (see **Figure 8**) have been recorded in the channel Facebook HCI-Collab [16], so the user can watch, download, and share the webinar online.

5.1 Collaborative applications

Adobe Connect [17] is the platform to use to transmit HCI webinars given their availability at no cost. It also allows management space for coordinators, moderators, and the speakers themselves. The platform allows you to share files and has screen views and also a chat. In addition, the number of participants is not limited, so the scope of the webinar may be greater. At the same time, you can choose between doing a public conference and doing it only for a specific group, something that can be interesting. **Figure 9** shows an initial section of HCI webinar about feelings analysis in educational contexts by Benemerita Universidad Autonoma de Puebla ("BUAP") researchers from Puebla, Mexico.

In addition, conducting webinars means presenting webinars in which communication can be in both directions, where communication is first from a speaker and then remote assistants can participate via the Internet. As in any conference, in the

Figure 8.
Facebook and HCI collaboration showing several HCI webinars during the year 2019 [16].

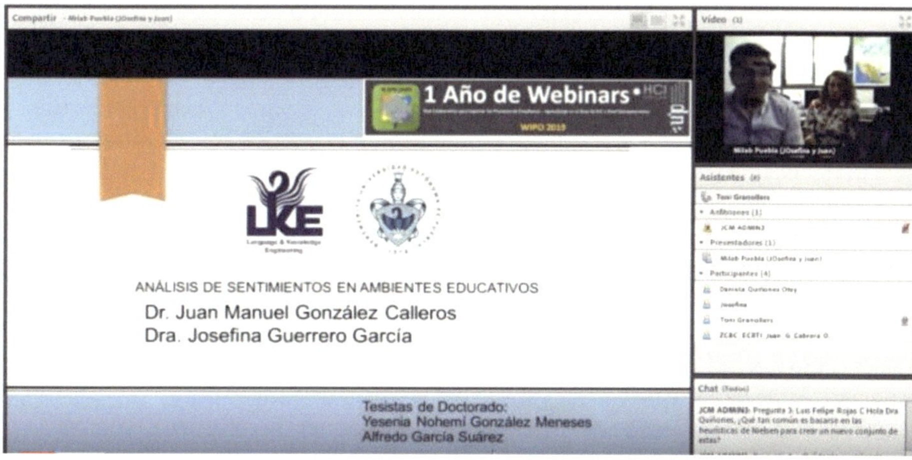

Figure 9.
Initial section of HCI webinar about feelings analysis in educational contexts using the Adobe Connect [17].

webinars the user can use a visual presentation (e.g., a PPT presentation), an audio presentation, and/or also a video from the presenters. Throughout the conference, they can communicate through an open chat window to ask questions to the speakers. The resources that the speaker can provide, such as documents, simulations and applications, among others, may also be available during and after the webinar.

Figure 10 shows examples of avatars about feelings in educational contexts by BUAP researchers. With Adobe Connect, it is possible to share it through the direct URL or insert it in a blog post. The user can invite others from the contact list; but in any case, they will be notified automatically when a public webinar is created.

A planning phase was developed in order to ensure and synchronize the webinar's organization efforts; it will be necessary to carry out a series of tests, as well as agree on the date, the start, the duration, and the topic of interest, HCI (one or two topics maximum). Once all this previous information has been determined, in the second phase (design phase), it is communicated by mail and/or by social networks on topics interested in HCI in Spain and Latin America such as the AIPO network and the HCI-Collab itself. It is in the third phase (construct phase) where the speaker carries out his presentation about HCI through the webinar platform; a moderator will be aware at that time not only to facilitate the activities of the presenter but also

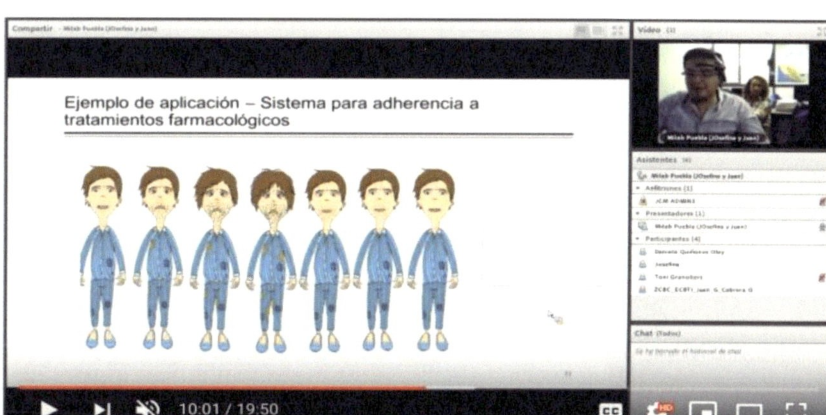

Figure 10.
Avatars of HCI webinar about feelings analysis in educational contexts of Adobe Connect [17].

to attend to the management of activities of the online audience, such as confirming communication to attendees and recording your concerns, questions, and comments. Attending to the participants' questions is important because if several have the same question or concern, it can be resolved at once and live. In turn, this will cause the participants to provide feedback to the comments and questions of the other attendees. This avoids massive emails and wasted time answering similar questions. In the evaluation phase, the academic resources that accompany each webinar, such as practices, interactive applications, transparencies, and the conference itself, will be available through the online repository of the HCI-Collab network. Other resources can make up the collection of the repository such as conversation forums, topics of interest, and collaborations that are based on the theme of the given webinar. It is important to indicate that attendees can contribute and share their own experience by participating as a speaker in one of the webinars to be carried out.

For example, the Facebook HCI-Collab video channel (see **Figure 8**) allows the recording of each seminar for later publication, thus speeding up the online video production process of an HCI theme to be available immediately at the end of each seminar. Each time a webinar is carried out, it requires covering a process that generally helps to organize the participants and display the presentation and distribution of the learning resources in HCI. Thus, the process is composed of four stages according to the proposed process model in **Figure 2**.

5.2 Services

The technology of service-oriented architecture [12, 13] was also exploited to identify several web services to support the main process of collaborative production of HCI webinars such as HCI webinar production and HCI webinar management (see **Figure 11**). The web services for the first process are as follows: prepare content for HCI webinar, make previous tests for an HCI webinar, give conference of HCI webinar, and answer questions from the audience. The second process, HCI webinar management, is supported by the web services: call for a new HCI webinar, accept an HCI webinar speaker, identify the speaker, introduce the speaker, get questions from audience, and publish a new HCI webinar.

In this case study, the collaborative production of educational resources is concreted in the repository of HCI webinar from a website HCI-Collab. In addition, the participants in the HCI webinars become members of HCI-Collab community [15].

5.3 Repositories

The initiative called "One year of HCI webinars" [15] helps to produce HCI educational resources in order to obtain a repository as extensive as possible online and free of charge on various topics related to HCI in Spanish. They are short-lived webinars under many topics. Something that we have been doing regularly, that is natural to us. The webinar has as its main interest topic, teaching HCI, mainly in the university field; all initiatives offered by companies, or other entities or groups, are also welcome.

The digital resources, as well as the experiences and successful projects presented in the webinar, can be capitalized under a platform as a collaborative network website to support the teaching-learning processes in the area of human-computer interaction at the Ibero-American level, HCI-Collab, a website that helps meet training and updating needs in the HCI area in the region (See **Figure 12**). The proposal was feasible in the event that each of the members of the HCI-Collab community reaches at least one webinar, during a year; a large amount of material was obtained with a fairly low collaborative effort. This repository represents a training corpus that can be used to complement HCI teaching.

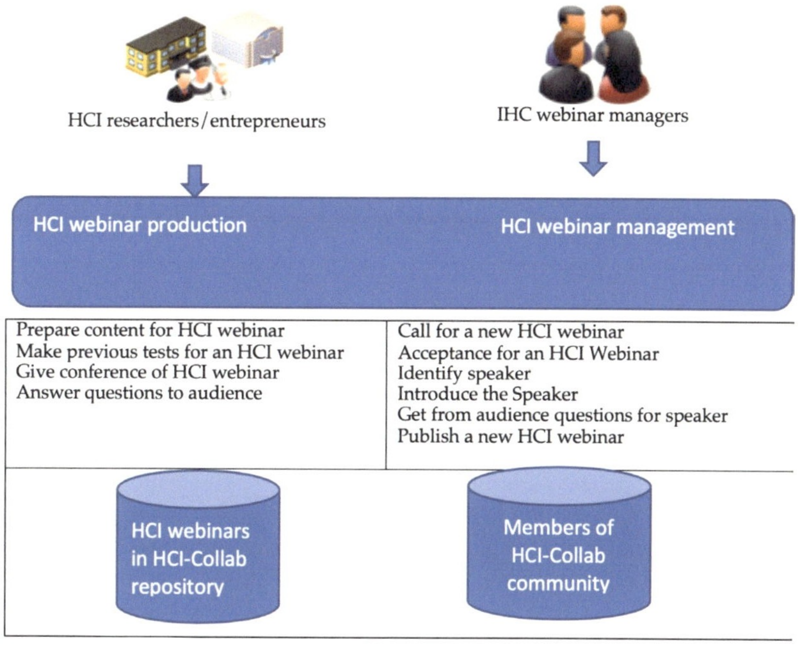

Figure 11.
Web services for collaborative production of HCI webinars.

Figure 12.
Website for the initiative One year of HCI webinars [15].

In addition, the Facebook HCI-Collab allows also the recording of each seminar for later publication, thus speeding up the online video production process on an HCI theme to be available immediately at the end of each seminar (see **Figure 8**). Each webinar can be shared through the direct URL or inserted into a blog post, and specific contacts can also be invited; but in any case they will be notified automatically when a public webinar is created.

6. Conclusion

This work proposes the use of a model for the collaborative production of online educational content in the area of human-computer interaction. The model encourages the multidisciplinary collaboration to produce and share the use of online HCI educational resources. The proposed model has been applied first to develop a textbook in HCI by several professors from universities in Latin America. The proposed model was applied for the collaborative production of open textbooks, considering the use of online services for the creation, adaptation, and mix and reuse of open textbooks. In addition, it considers strategies for the implementation and adoption of the initiative for such textbooks. This model was implemented also for educational resources in terms of a series of webinars, which are offered to the Latin American community as a means of obtaining and accessing learning and reference resources in Spanish in the discipline of human-computer interaction. The initiative is proposed as a strategy to promote knowledge about various HCI issues, which will be presented by renowned Latin American researchers and entrepreneurs. It is expected that this initiative will strengthen the HCI-Collab network and allow the entry of new members. In itself, the objective of the initiative to be achieved is twofold because, on the one hand, it is intended to contribute to the generation of educational resources in the HCI area and, on the other hand, to help in the formation of communication mechanisms between professionals and academics. As a final conclusion, it can be said that carrying out the HCI webinars for a year will favor new forms of organization and participation, allowing the emergence of new collaboration scenarios in which you can attract, manage learning, and increase knowledge, which they have added value for professionals and organizations.

Several subjects to study are considered here as future work, in particular the study of user satisfaction and the development of better services [18], to access the HCI educational content to cover a major number of communities in Latin America.

Author details

Jaime Muñoz-Arteaga
Universidad Autónoma de Aguascalientes, Aguascalientes, México

*Address all correspondence to: jaime.munoz@edu.uaa.mx

References

[1] Dix A. Human-Computer Interaction. 2nd ed. USA: Pearson Prentice Hall; 2003

[2] Muñoz-Arteaga J, González JM, Sánchez A. Interacción Humano-Computadora en México, Pearson Editorial, 2015. ISBN 978-607-32-2920-9

[3] Preece J, Rogers Y, Sharp H. Interaction Design: Beyond Human-Computer Interaction. 4th ed. Chichester: John Wiley & Sons; 2015

[4] Collazos C, Ortega M, Granollers T, Rusu C, Gutierrez F. Human-computer interaction in Iberoamerica academic, research, and professional issues. IT Professional. 2016;**19**(5):14-17

[5] Golder P. A new architecture for a collaborative authoring system. Computer Supported Cooperative Work. 1993;**2**:159-174. Kluwer Academic Publishers

[6] Ochoa X, Muñoz-Arteaga J, Casali A, Frango I, et al. LATIn Project -Latin America open Textbook Iniciative, ("Iniciativa Latino Americana de libros de texto abiertos"). Alfa III (DCI-ALA/19.09.01/11/21526/279-155/ALFA III(2011)-52)

[7] Henderson S, Nelson, D. The Promise of Open Access Text-Books: A Model for Success. Tallahassee, Florida, USA: Found for the Improvement of Postsecondary Education (FIPSE); 2011

[8] Carmenado I, Figueroa B, Gómez F. Methodological proposal for teamwork evaluation in the field of project management training. Procedia-Social and Behavioral Sciences. 2012;**46**:1664-1672. DOI: 10.1016/j.sbspro.2012.05.358

[9] Giving H. Knowledge for Free: The Emergence of Open Educational Resources. Paris, France: OECD Publishing; 2007

[10] Boehm B. A spiral model of software development and enhancement. IEEE Computer. 1988;**21**(5):61-72

[11] Muñoz-Arteaga J, Hernandez Y, et al. "Temas de diseño en Interacción Humano-Computadora." Available from: https://www.researchgate.net/publication/322215720_Temas_de_diseno_en_Interaccion_Humano-Computadora, Creative Commons. [Accessed: 14 September 2019]

[12] Luttikhuizen R, Dikmans L. SOA Made Simple. USA: Packt Publishing; 2012. ISBN: 9781849684163

[13] Thomas E. Service-Oriented Architecture (Paperback): Concepts, Technology, and Design. USA: Prentice Hall; 2016

[14] LACLO, Latin American Community of Learning Objects. Available from: https://laclo2019.com/. [Accessed: 14 September 2019]

[15] Granollers T, Collazos C, Muñoz-Arteaga J, HCI Webinar in HCI-Collab. Available from: http://hci-collab.com/. [Accessed: 14 February 2019]

[16] Granollers T, Collazos C, Luna H, HCI Webinar in Facebook. Available from: https://www.facebook.com/pg/hcicollab/events/. [Accessed: 14 September 2019]

[17] Adobe Connect. Available from: https://helpx.adobe.com/support/connect.html. [Accessed: 14 February 2019]

[18] Chang E, West M. Digital ecosystems: A next generation of the collaborative environment. In: Proceedings of the Eighth International Conference on Information Integration and Web-based Appli-cations & Services. Yogyakarta, Indonesia; 2006